Silk Culture

Silk Culture

by
S.K. Ananthanarayanan

BIOTECH

2026

Biotech Books
Delhi - 110 035

ISBN10 81-7622-189-9
ISBN13 978-81-7622-189-4
EAN: 9788176221894

Published by : **BIOTECH BOOKS**
4762-63/23, Ansari Road,Darya Ganj,
New Delhi- 110002
Phone : 011-43003222

e-mail: biotechbooks@yahoo.co.in

PRINTED IN INDIA

Preface

Silk culture, or sericulture, is an art known to humankind for many millennia, perhaps even as early as 300 B.C. Widely ascribed as the invention of the Chinese, silk today is one of the most coveted yet expensive fabrics, and the silk industry, with great help from technology, and better awareness, books all set to usher in a new era of burgeoning growth and development.

This book traces the history and development of sericulture to the present times, when it has become an important industry for countries like India and China, and offers a great perspective on the industry's evolution. It also describes the sericultural process in some detail, and looks at the present state of the industry. Besides providing various statistical figures, it contains many industrial facts crucial to the knowledge of the sericulture farmer. It looks at the market forces of demand and supply of silk, cost of production and selling prices, the transition of the industry from a cottage industry to a large scale one, etc.

Being a comprehensive guide to the silk culture industry, it is hoped that this book proves to be highly useful for the reader interested in learning about the fabric and its developmental process.

Editor

Contents

1

Mulberry Trees

Botanists discovered only one native species of mulberry tree in North America, viz. the, red, (*morus rubra,*) which has an extensive range. Michaux assigns the same limits north to it as to the majestic and beautiful tulip tree, (*liriodendron tulipi fera,*) viz. the northern extremity of lake Champlain; but it also grows in Massachusetts. Southward and westward, it abounds in all the States, and has been recently found as far west as the lower part of the river Canadian.

The leaves of the red mulberry tree are large, generally entire, but sometimes divided into two or three lobes, rounded, heart shaped, stud denticulated, of a dark green colour, thick texture, rough and uneven surface. The sexes are usually separate, though sometimes they are found upon the same tree. The fruit is of a deep red colour, an oblong shape, and of an agreeable, acidulous sugary taste; it is composed of the union of a great number of small berries, each of which contains a minute seed.

The tree often exceeds sixty feet in height, and two feet in diameter. The wood is of a yellowish hue, approaching to lemon colour, is fine grained and compact, and when perfectly seasoned, is as durable as the white locust, on which account it is highly esteemed for posts, and, by ship and boat builders, for the upper and lower frames of vessels,, knees, and floor timbers, and the ribs of boats. In Tennessee, and probably in the other Western States, when a native forest is cut down, if the land be enclosed, a growth of red mulberry trees soon takes its place.

There are several varieties in the red mulberry tree, depending on the leaves and fruit

1. Leaves all orbiculated, (round.)
2. do, deeply lobed.
3. do, with three short lobes.
4. Fruit, borries nearly white.
5. do. do. blueish purple.
6. do. do. red and long.
7. do. do. blackish red.

Mr. C. C. Robin, a French traveller, mentions a species with leaves similar to the red mulberry in shape, but rough and shining on the upper surface, and downy underneath, with white fruit, and growing in Louisiana. That the leaves of the native red mulberry tree agree perfectly with silk-worms, and yield very good silk, is a fact so well established by the experience of more than a century, that, to doubt it would amount to an absurdity. It appears, however, that the leaves do not suit the constitution of French worms, and the author, of the experiment, which he made to satisfy himself, on this point, therefore, decides against the fitness of the tree for the food of the insects. He acknowledges that they eat the leaves of both the white and red species indiscriminately, when mixed, without evincing a preference for either, and that when they were placed in separate parcels within half an inch of each other, they never left the latter until they were all devoured.

From the character of the leaves and fruit of a native mulberry tree growing in Washita, there is reason to believe that it is a different species from the *Morus Rubra*. The leaves received, are three-lobed, three-nerved, unequally serrated; base, subcordate entire; lobes, ovate oblong acute, or acuminate; sinuses, broad, with large interjected acute teeth. Both surfaces rough.

The leaves are larger than the red species: tipper lobes more ovate, with base narrowed; no pubescence beneath: lateral lobes narrower than the middle. Teeth of the sinuses, sometimes entire, sometime with a few unequal teeth on the side. Upper sinuses broader than the lower.

White Mulberry Tree

The white mulberry tree—Mores Alba, is a native of Asia, and was introduced into Italy by some of the survivors of the last crusade. Olivier do Serves relates, that the French who accompanied Charles the Eighth, in his invasion of Italy, in 1494, struck with the abundance of the trees in that country, and with the profit derived from the culture of silk, determined to introduce it into France. This was done by Guy Pape, of Saint Auban, after the peace, who planted it near Montelimart. In I802, Faujas de Saint Fond saw the original tree, around which Mr. Lachaux, to evince his respect for this monument of agriculture, and parent of all the white mulberry trees in France, hall built a wall.

There are several species of this tree, and numerous varieties, the result of cultivation, soil, climate, and the play of nature. The forms of the leaves are extremely variable. Mr. Audibert, an experienced cultivator in

France, says, " that the same tree will have leaves divided into several lobes, when young; and, when it becomes old, they will be entire. Others have the second crop of leaves differently formed froth the first: some again have entire leaves in the Spring, and lobed leaves in the Autumn. Hence, it is extremely difficult to assign positive character; to the different varieties, particularly when they slew no diversity in appearance, except in the shape; or the leaves."

All the practical writers oil this subject agree, that the proper soils for the mulberry tree are dry, sandy, or stony; the more stony the better, provided the roots of the trees can penetrate among them. The situation should be high; low, rich, and moist land, never produce nourishing leaves, however vigorous the trees may grow. They are always found to be too watery.

The modes of propagation, are,

1. by seed;
2. grafting;
3. budding;
4. layers;
5. cuttings;
6. suckers.

The ripe fruit may be sown in drills in ground previously prepared; or the seeds may be washed out of the pulp, and mixed with an equal quantity of sand or fine mould, and then sown. They should be covered about a quarter or half air inch deep. The, seeds will soon vegetate if the ground be rich, and will live through the Winter, unless the cold should be unusually severe. In that case, they should be covered with straw or long manure: in the course of the next season thin the plants, so that they may be a foot apart. Seeds intended to be sown in the Spring, or to be kept, should be washed out: for they are apt to heat or mould if allowed to remain in the fruit.

Every tiller of land knows the fertilizing effects of frost and snow; the land, therefore, destined, for the Spring sowing, should be dug or ploughed in the preceding Autumn, left rough all Winter, and harrowed or raked fine as soon as the season will permit, and the seed sown in drills. The young plants must be watered in dry weather, and weeds carefully kept down.

Weeds will not only stint the growth of the plants, but cause diseases in them, which may affect the future vigour and health of the tree. When a year old, some of them will be fit to plant in nursery rows; the same plants may remain in the seed bed a second year, and then be transplanted: the plants at two feet distance from one another, that there may be room for cleansing and dressing the ground.

At transplanting, cut off some of the roots, especially those that are ragged or decayed, and the tap root, to force out lateral roots; and also the tops; at six or seven inches from the ground. In France, they transplant just after the fall of the leaf in the Autumn. When the plants in the nursery are sprung, strip off the side buds, and leave none but such as are necessary to form the head of the tree. The buds which are left, should be opposite to one another. If the plants in the nursery do not shoot well the first year, in the month of March following cut them over, about seven inches from the ground; this will make them grow briskly. They should also be watered with diluted barn-yard water.

When the plants are grown to the size of one inch in diameter, plant them out in the fields where they are to remain; make the holes six feet square, and dress the ground two feet round the plants, trim the roots, and press the earth on the roots as the holes are filled. During, the first of planting out, leave all the buds which the young trees have pushed out on the top, till the following Spring, when none are to be left but three or four branches to form the head of the tree. The buds on these branches should be on the outside of them, that, the shoots may describe a circle round the stem, and that the interior of the tree may be kept open; and as the buds come out, take off all those which appear upon the body of the tree.

For several years after, every spring, open the heads of the trees when too thick of, wood, and cut off any branch which crosses or takes the lead of the rest; leaving two buds on the out side of every trimmed branch. Count Verri, an experienced cultivator of the mulberry tree, recommends to leave only two buds at the end of each branch-preferring those which are outside, and opposite to each other; and when three buds appear together, to leave the middle one, which is always most vigorous, and to detach the two on each side of it.

If the superior buds do not push well, the two next lower ones, only, must be left. Every farmer and nurseryman, knows the very great importance of dressing the ground round young trees, twice in the course of the year, and of planting stakes by them, to ensure an upright, straight growth, and to prevent their being shaken by wind, or levelled by storms.

Grafting and budding may be performed on the mulberry tree in the modes usually adopted for other trees: So great is the prejudice in Italy in favour of grafting the mulberry trees, that Dandolo says, "even the wedge mulberry trees are grafted." The reason for this is, that the grafted tree yields more leaves than those produced from seed; but he decides, as will be hereafter seen, in favour of the latter. Nevertheless, as it may be wished to propagate a particular species, or variety of mulberry tree, the leaves of which are found to produce better silk than common, the process may be adopted which is generally preferred in France for grafting the mulberry and large chesnut, and is as follows:

A branch must be selected, sound, and of the growth of the preceding year, when possible, and, while the sap is running: this is to be cut off some inches from the trunk, or further off, according to its strength or size. Then, without loss of time, it is to be slid upon the wood, until its lower extremity touches the basis of the strips. If the cylinder applied to the wood is proportioned to it, and if it covers all the wood, and unites exactly with it, the strips are to be cut off below the cylinder, by a circular cut; and, after having made the two barks join, the united parts are to be covered with a mixture of turpentine, wax, and rosin, spread on thin leather.

When it is difficult to find a branch, the bark of which will fit exactly the branch which is cut in strips, the following expedient may be adopted: If it be too narrow, a branch must be slit longitudinally, on the side opposite to the bud, and the cylinder being taken oft, is to be applied to the wood: then a part of a strip is to be raised, that corresponds to the deficiency in the cylinder, and must be divided in its whole length, of a breadth wanting in the ring, so that this division fills the vacant space: finally, raise up and surround it with the strips as already directed. If the bark cylinder be too large, cut it to the diameter of the wood, then fit the edges as nearly as possible, and cover them with the strips.

Laying—The propagation of trees by layers is only necessary for such trees as do not grow freely by cuttings, or of which seeds cannot be easily procured; but as neither of these reasons apply to the mulberry tree, no directions on the subject are necessary.

Cuttings—These should be taken from perpendicular shoots, and particularly those that terminate branches: these will certainly produce straight and handsome trees; while cuttings taken from horizontal branches, will ever have an inclination to grow in a spreading manner. The cuttings should be of the last Summer's shoots, and from six to fifteen inches in length. Plant them in rows, in shady borders, early in the Spring, and about two-thirds their length in the ground; close the earth well about them, and in dry weather let them be watered. After a year, they may be transplanted into open nursery rows, if well rooted.

Suckers—These may be separated from the parent plants, early in the Spring, each with some roots, and planted either in nursery rows for a year or two, or the largest in the place intended for them, permanently. It is essential for the success and quick growth of young plants, that the ground around them should be fine, so that the tender roots may not be obstructed in their progress. SLUGS and SNAILs are very destructive to mulberry plants, and will eat down numbers of them in one night, and, in a moist season, will ruin a nursery, if not prevented. To prevent their depredations, surround the bells with soot, hot lime, or ashes, sprinkling a fresh parcel when they have been wet with rain; but no soot should be put on the beds, being too acrid for the plants. If the insects appear, destroy them after sunset.

Mr. Pullein says the best defence against them is a hair rope, which should be trimmed so as to be made as bristly as possible: this being pinned to the ground round the border of the bed, will so prick their tender skins that they would not venture to cross it. In order to secure early food for silk-worms, the cultivator should plant a hedge of mulberry trees, or a few young trees in a warm situation. In the Spring they should be covered with straw, or arty other tray, at night, to prevent injury from frost. As the white mulberry tree grows readily, with proper care, they may be easily multiplied. They may he planted along the fences of a farm, and if the branches be, wattled through the rails, they will form an impenetrable hedge, and the fence will never require renewing. A great advantage will attend such a hedge, viz: the ease with which the leaves can be obtained: the avoidance of falls by young persons front large trees, is another consideration in favour of propagating mulberry hedges.

This accident play also be prevented by annually pruning the trees, and retraining their upright growth to the extent they are commonly permitted to attain. The quantity and quality of the leaves will also be thus, greatly increased. Practical writer, particularly Sauvage, dwell upon the importance of annually pruning trees, for these reasons; and a gentleman, recently returned from all extensive tour through Europe, and the East, states, that the cultivators of sill:, particularly in Syria, were unanimous as to the superiority of the silk produced by worms fell on the leaves of trees annually pruned, over that made by worms nourished by trees, the growth of which was unrestrained. The American cultivator is urged to attend to this hint, whether he feeds silk-worms oil the native or foreign tree.

The white mulberry trees which have been planted in Massachusetts and New Hampshire, grow well, and are not injured fly frost. They have trot yet been introduced into Maine, but there is no reason to doubt their flourishing in that State, and also in Vermont. Should, however, the cold prove injurious to them, the Tartarian mulberry, which is cultivated for silk-worms in the silk-rearing province of Zacsyn, in Russia, might be easily introduced.

There are several varieties of the white mulberry tree, two of which have white berries; one red, and the other black. There are two varieties in the leaf; one has a leaf deeply indented, and supposed to contain little nourishment; the other larger, and not much indented, or lobed. The common grafted mulberry is a variety of the first of these two, and itself comprises the following varieties: 1st. with white berries; 2d. with red berries; 3d. with black berries; 4th. with a large leaf, called the mulberry of Tuscany; 5th. with a middle sized leaf, dark green, called *faglia giazzola*; 6th. with a small leaf, of a dark colour, ratherthick, called double. leaf, more difficult to pick, and the best calculatedfor silk-worms.

Besides these, there are numerous other species and varieties in Europe, the result of cultivation, but the detail oft heir names, and their description, would be useless. A late French writer, and a practical man,t after having taken tip several pages of his work in describing them, decides in favor of those called Colombasse and Colombassette, as being favorable to the health of the silk worms, and as yielding the greatest quantity and the best quality of silk. The leaves are small, thin, light, and silky. The fruit, when at full maturity, is yellowish, and very large. He also approves of the species called the rose mulberry; the leaves of which are larger, and of a deeper green than the preceding species: the fruit is reddish, and as large as that of the other species.

Thereare five different substances in the mulberry leaves:

— the solid or fibrous substance;

— the colouring matter;

— water;

— the saccharine substance; 5th. the resinous substance.

The saccharine matter is that which flourishes the insect, and forms its animal substance. The resinous substance is that which, separating gradually from the leaf, attracted by the animal organization, fills the two silk vessels. According to the different proportions of the .elements which compose the leaf, it follows, that cases may occur, in which a greater weight of leaf, may yield less that is useful to the silk-worm, as well for its nourishment, as with respect to the quantity of silk, obtained from the animal. Thus the leaf of the black mulberry (a native of Europe) produces abundant silk, and strong, but coarse.

The leaves of white mulberry trees, in high lands, exposed to cold dry winds, and in a light soil, produce a large quantity of strung silk, of the purest and finest quality. The leaf of the same tree planted in damp situations, and in a stiff soil, produces less silk, and of a duality less pure and fine; but the chief cause which influences the fineness of the silk, is the degree of temperature in which the silk-worm is reared, as it is hoped, will be demonstrated in the course of this work.

According to Dandolo, the leaves of the broad-leaved white mulberry, are little nutritious. The next has a middle size leaf, thick and of a dark green colour. The best mulberry leaf cf any species, is that which is called the double leaf; it is small, not very succulent, of a dark green colour, shining, and contains little water; which may be easily ascertained, by drying some of them:the tree produces them in great abundance.

And old tree produces better leaves than young trees; as the tree grows older, the leaves diminish in size. The leaf of the wild [seedling] tree, contains the greatest proportion of both the nutritive and silky substances. This opinion of Dandolo, in favour of the wild mulberry tree in preference to the grafted tree, as food for silk-worms, is thus supported: he ascertained the following facts:

1. That 141/2 lbs. of wild mulberry leaves, will produce a pound and a half of cocoons; while 201 Ibs. of the leaves of the grafted mulberry, are required to yield the same quantity.
2. That 741/2 lbs. of cocoons, proceeding from silk-worms, fed an leaves of wild mulberry, give about 14 oz. of very tine silk; whilst generally the same weight of Silk-worms, fed with leaves of the grafted mulberry, only yield eleven or twelve ounces of silk.
3. That the silk worms fed oil the wild leaves, are always brisker, and have better appetites.

The result is, that of taking two trees of equal age and vigour, the grafted tree yields 50 lbs. of leaves, and the wild tree only 50 lbs.; the weight of nutritious substance will be nearly equal in each. To this authority may be added that of Mr. Martleroy, an experienced cultivator in France, who found that silk-worms, fed with the leitics of the seedling mulberry tree, were more healthy, vigorous, and less subject to diseases, than those which are fed upon the leave of the grafted trees.

It is impossible to ascertain with any degree of tolerable accuracy, the quantity of leaves which an acre, or even a single mulberry tree will produce, owing to the operation of one or more of the numerous causes influencing their production: such as the particular species of trees; the greater or less care taken of them, in their early stages of growth; their position, and distance from one another, when planted out; the soil, and especially bad or injudicious pruning, or total neglect of this necessary measure.

The estimates are accordingly, very various. Mr. Fitch says, that one acre of full grown trees, set one and a half' rods apart, will produce 40 lbs. of silk. One tree yielded food for worms, which gave four pounds of silk. Mr. Tufts confirms this calculation, but does not state the distance at which the trees stood from one another. Mr. Storrs says, that 14 a full grown tree will feed 6,000 worms, which will produce one and a half pounds of silk.

An acre of trees will produce 60 Ibs. of raw silk in one season." Mr. R. Falley, now of Ohio, had 18 lbs. of silk from about 100 trees, part of which were young, in Massachusetts. In the year 1188, nearly four pounds, were produced from seven trees, and one pound from eight trees, eight years old from the seed. According to Dandolo, it is a certain fact, that, if silk-worms are well managed, 21 lbs. of mulberry leaves will be sufficient to obtain a pound and a half of cocoons.

Twenty pounds four ounces, will feed worms enough, to produce a pound and a half of cocoons. In Dalmatia, he obtained one pound and a half of cocoons from 15 lbs. of leaves, which yielded one pound and a half of silk: ninety-seven pounds eight ounces of leaves, will produce seven and a half pounds of cocoons. Miss Rhodes could scarcely support 10,000 worms, on the leaves of twelve large trees in England.

According to Lambruschini, 100 lbs. of clean leaves will feed worms which will produce six pounds of silk. Dandolo has prescribed the proportions of leaves which will be required for five ounces, and one ounce of silkworms in their progressive stages, and lays down the weight of food which will yield certain quantities of cocoons, produced by different species of the insect: as he writes from ample experience, his directions and conclusions may be safely relied on. He recommends that the trees when transplanted, be not stripped for three years, and to thin and prune the branches in the fourth year. In the fifth year, they may be stripped without danger; but he advises cultivators to permit a tree to rest one year after being stripped, in order to recover froth the loss of its leaves.

This reasonable caution should be attended to particularly with young trees; for leaves are considered as the lungs of a plant or tree, and mainly contribute, by the absorption of moisture, and the principle of vegetation from the atmosphere, to promote their growth and vigour. If deprived of these organs, annually their growth will be greatly impeded the people of Connecticut and Massachusetts are of opinion, that trees only two years old, may be stripped without injury, provided the leaves near the ends of the branches are left, and the main stem be dot touched. Count Verri recommends to pass the bands from the lower part of a branch to the top, and to pick the leaves in a bag, the mouth of which is to be kept open by a hoop, with a hook to suspend it. As practical rules, these directions are worth attention; but in observing the last, it is essential that the leaves be not bruised.

The experiment was tried in Connecticut, of feeding worms with young mulberry plants, the seeds of which were sown broadcast, and the plants mown as wanted. The roots soon died. As the leaves of such plants contain so much less nourishment than those of old trees, the measure cannot be recommended, except for early food.

Mulberry Hedge

Choose grafted mulberry plants of one year old, and place them eighteen inches apart, in a furrow prepared some months before. To these may be added grafted plants of two years old, which have grown well, and which have been cut off to increase their vigour: but they must be separated from the others, to prevent irregularity in the hedge. Cut them at four or six inches from the ground, leaving two buds opposite each other: remove all the rest. In this way, the stalk has two vigorous branches the first year. In the spring season, cut one of these two branches on the same side, at about one foot from the ground, so that each plant has a short and a long one. Bend, horizontally on the same side also, one after another, all the preserved, branches, and fix them with willow-withs, so that they may form a line parallel to the earth, and leave untouched the entire branches.

Experience has taught, that, if their extremities be cut, they grow slowly, and even die. At the commencement of the third year, the plants will have branches to form a hedge. Then cut them about two feet from the ground, but do not use the leaves. This plan has the following advantages

1. The mulberry plants being grafted, and well arranged, and those of the first year being separated from those of the second, the hedge will grow with force and regularity.
2. The plants which have been cut down, will push out near to the ground, and furnish numerous branches. When a plant dies, replace it by layers front an adjoining one; if a new plant be substituted, it rarely succeeds. The leaves from the layers must not be pulled during the three first years.

M. Bonafous says, that the leaves of the paper mulberry tree agree perfectly well with siik-worms in their fifth age. The character of this writer, and his great personal experience, entitle him to full confidence, and their can be no reason to doubt that the worms to which he gave these leaves, eat them without injury; nevertheless, other silk-worms were less fortunate. M. Deslonchamps says, "that, of 100 silk-worms to which he gave nothing but the leaves of the paper mulberry tree, 92 died; and the remaining Height, which survived the rest thirty-six days, would also certainly have died, if; at this time, he had not substituted the leaves of the white mulberry.

This new food enabled them to live twenty and twenty-two clays longer, and to spin their cocoons; but these were so small, that eight of them did not weigh more than two ordinary cocoons." From the above statement, it appears that M. D. gave the leaves of the paper mulberry to young worms; whereas Mr. B. confines their use to the fifth age—a time when they have acquired their full growth, and their appetite is voracious. This variation in the treatment may readily account for the different results of the two cultivators. The experiment with the paper mulberry leaves, in the fifth age, is worth trying with 50 or 100 worms: and, if made, it is requested that the result may be communicated to the public.

The black mulberry tree having been mentioned as suitable for the Southern States, it may be proper to say something on the subject. The leaves of this tree, it appears, are preferred in Spain, where superior silk is made. They are said to be much more nutritious than those of the white mulberry; but a late practical writert denies this position: for, in the year 1823, he fed some silk-worms on the leaves of both these trees; and the cocoons of 100 of them which had eaten the black, weighed from two to three grains less, than others of another 100 worms which fed on the white.

Mr. Nysten also says, that it is late in attaining its full growth, and difficult to propagate, either by grafting or by sowing the seeds. The leaves also come out full ten days later than the white. This is a serious objection, as early food for the worms is of primary importance: finally, the black is comparatively a tender plant. For these reasons, it cannot compete with the native red, or the foreign white mulberry tree.

2

Silk Culture

The culture of silkworms (*Bombyx mori*) began in China about 5000 years ago. The ancient Chinese were the first to discover that the cocoons of a lowly caterpillar could be unwound and the filaments woven into some of the most beautiful fabrics in the world. Silk was so highly prized that the lustrous fabrics were reserved exclusively for the royalty of China.

The secrets of silk production were closely guarded by the Chinese for thousands of years. It was not until about 300A.D., that sericulture spread to Korea and from there to Japan. In 552 A.D. two Nestorian monks, under orders from the Emperor Justinian, smuggled silkworm eggs from China and bought them to Europe. This was the beginning of the silk industry in the West. Sericulture spread rapidly throughout Europe and Asia during the following centuries.

The English began making silk in the 13th century, but their success was limited by their cool and damp climate. Silk culture began in America in 1603 when silkworm eggs and mulberry seeds were sent to Virginia by order of King James !

The English hoped to rival the French and Italians in silk production by having the American colonists raise silk for them. Silk was produced erratically in Virginia and Georgia until about the 1760's. By this time, cotton and tobacco had proven to be more lucrative products for the southern colonies.

About 1760, Nathan Aspinwall, a native of Mansfield who had a mulberry orchard on Long Island, began to promote silk culture in Mansfield and nearby towns. He planted an orchard in town and in 1762, he brought the first silkworm eggs to Mansfield. With the aid of Dr. Ezra Stiles of Yale, he persuaded the legislature to pass an act which paid ten shillings bounty for every one hundred trees planted and three pence per ounce for raw silk. The legislature further encouraged the culture of silk by sending a half ounce of mulberry seed to every parish in Connecticut.

Silkworms were raised throughout the state in the late 1700's but only in Mansfield and some of the surrounding towns did the home industry continue much beyond the turn of the century.

By 1800 three quarters of the people in Mansfield were raising silkworms in their homes. The men probably planted the trees, then women and children gathered the leaves, cared for the worms and reeled and spun the raw silk. A spare room, an attic, a barn loft or an out building was used as a cocoonery.

In 1830, the Chinese Mulberry (*Morus multicaulis*) was introduced from China. It was highly touted as superior to the White Mulberry (*Morus alba*), shown to the right in the photo. It grew rapidly, was easily propagated from cuttings and silkworms fed upon its leaves produced higher quality silk.

The culture of *M. multicaulis* was promoted by Congress and soon growers up and down the Atlantic Coast were caught up in a wild speculation that rivaled the "Tulip Bulb Craze" in 17th century Holland. Prices rose rapidly as people rushed to invest in the new mulberry variety. In 1834, one year old trees sold for $3 to $5 per hundred. Soon the price rose to $500 per hundred and ultimately two trees sold in North Windham for $100 a piece.

The value of the trees was soon greater than that of the silk that could be obtained from them. Around 1840, the market for *M. multicaulis* crashed and many fortunes were lost. Disillusioned farmers tore out their mulberry orchards and the whole business of silk culture fell into disfavor.

Silk culture remained a profitable home industry in Mansfield until about 1844 when a blight devastated mulberry orchards throughout the country. The blight coupled with the crash of the *M. multicaulis* market led to the demise of silk culture in America. Few continued to raise silkworms after mid-century and those who did found it difficult to compete against foreign countries like China where labour was cheap and plentiful. The local silk factories, which were well established by this time, continued to operate using imported raw silk.

Home Production

In the early days of Mansfield's silk industry, the silk cocoons were unwound in the home using a simple hand reel – the same as was also used for measuring skeins of woolen yarn or linen thread. The cocoons were placed in a kettle of hot water and then were stirred with a brush. The silk fibers which adhered to the brush were then attached to the reel. Usually about ten cocoons were unwound (reeled) at a time. The fibers were pressed together to form a single strong thread which could then be easily wound as the reel turned.

The Hanks Silk Mill was built in 1810 by Horace and Rodney Hanks on the brook near 198 Hanks Hill Road. Using water from the brook to propel their machinery, they attempted to make sewing silk and silk twist. The mill was not a commercial success due to the crudeness of the machinery. However the mill is of great significance as the first in America to utilize water power. This was the first step towards mechanization in the American silk industry.

Recognizing the significance of the Hanks Silk Mill, Henry Ford purchased the building in 1930. It was reconstructed in his Greenfield Village Museum in Dearborn, Michigan, where it can be seen today.

Successful Silk Processing

In 1827 or 1828, a young throwster named Edmund Golding came to America from Macclesfield, England. He arrived in Mansfield, which was then considered the center of the American silk industry, expecting to find ready employment. He was disappointed, however, to find that the American silk industry was far behind that of England. The doubling and twisting operation he was trained in was here crudely done on the hand spinning wheel.

Alfred Lilly took an interest in Golding persuaded him to describe the machinery he had used in England. Believing that Golding's drawings were sufficient to produce workable machinery, he formed a partnership with Capt. Joseph Conant, William Fisk, William Atwood, Storrs Hovey and Jesse Bingham. The company was incorporated under the name Mansfield Silk Company in 1829.

The doubling and twisting machinery built under Golding's direction was an improvement over the accelerated spinning wheel used in the Hanks silk mill, but it was unable to operate using the crudely reeled local silk. At Golding's suggestion, the company began to import raw silk from England and with this they achieved moderate success.

The new mill attracted many visitors to Mansfield. Among them was Edmund Brown, another Englishman who had a silk tassel business in Boston and hoped to find a local source of raw silk. He was familiar with the English method of silk reeling and was able to direct the company in producing a more effective reel.

A number of improvements to the reeling process were suggested, including the French "Piedmont Reel" introduced to Mansfield in the 1830's. It was Nathan Rixford, however, who finally resolved the problems of reeling native silk. Rixford, who had just established himself as a builder of machinery in Mansfield Hollow, visited the Mansfield Silk Company's mill and examined its machinery. He then made considerable improvements upon it and in 1839 began to operate his own silk mill in Mansfield Hollow. With his improved machinery, he was able to produce, for the first time, reeled silk that was equal to the imported.

Historical Outlook

Historical evidence shows that silk was discovered in China and that the industry spread from there to other parts of the world.

Silk Discovery Secrete

The Chinese has used silk since the 27th century B.C.. During the Roman Empire, silk was sold for its weight in gold. The Chinese domesticated silk worms and fed them with mulberry leaves. They unwound the silkworms' cocoons to produce long strands of silk fiber.

Silk is one of the oldest fibers known to man. Its discovery as a weave able fiber is credited to the Lady Xi Ling Shi, the 14-year-old bride of the Emperor Huang Ti, the so-called 'Yellow Emperor'. One day in 2640BC, according to Confucius, she was sitting under a mulberry tree, drinking a cup of tea into which a silk cocoon fell from above. She noticed the delicate fibers start to unravel in the hot liquid and has been credited as the first person to 'reel' or unravel a silk cocoon and use the filament to create a yarn for weaving. Whether or not the legend holds true, it is certain that the earliest surviving references to silk production place it in China and that for nearly 3 millennia, the Chinese had a global monopoly on silk production.

The Chinese kept the secret of the beautiful and value added material that, they were producing from the rest of the world for more than 30 centuries. Travelers were searched thoroughly at border crossings and anyone caught trying to smuggle eggs, cocoons or silkworms out of the country were summarily executed. Demand for this exotic fabric eventually created the lucrative trade route now known as the 'Silk Road,'. of which mention is made as early as 300BC in the days of the Han Dynasty, taking silk westward and bringing gold, silver and wools to the East.

Silk was exported along the Silk Road (the ancient trade route linking China and the Roman Empire). This trade brought China a great wealth, but the Chinese did not give away the secret on how silk was produced.

The Chinese could not keep their monopoly forever. The industry is said to have spread to Tibet when a Chinese princess, carrying silkworm eggs and mulberry tree seeds in her headdress, married the king of Khotan in Tibet. From Tibet the industry spread slowly to India and Persia.

According to Western historians, mulberry-tree cultivation spread to India through Tibet during 140 BC and cultivation of mulberry trees, rearing of silkworms began in the areas flanking the Brahmaputra and Ganges rivers.

According to some Indian scholars silkworms (*Bombyx mori*) were first domesticated in the foothills of the Himalayas. Evidences in ancient Sanskrit literature reveals that certain kind of wild silks were cultivated in India from time immemorial.

When British came to India, the flourishing silk trade exploited and developed silk centres in many parts of the country. The Company exported large quantities of silk produced in West Bengal to England. The Company's monopoly was abolished in 1836 and the entire trade turned over to private enterprise, due to improper organized system the silk industry in West Bengal declined. By the time other silk producing states in the country viz., Jammu & Kashmir, Mysore have developed the industry.

Sericulture in India

Silk is a way of life in India. Over thousands of years, it has become an inseparable part of Indian culture and tradition. No ritual is complete without silk being used as a wear in some form or the other. Silk is the undisputed queen of textiles over the centuries. Silk provides much needed work in several developing and labour rich countries. Sericulture is a cottage industry par excellence. It is one of the most labour intensive sectors of the Indian economy combining both agriculture and industry, which provides for means of livelihood to a large section of the population i.e. mulberry cultivator, co-operative rearer, silkworm seed producer, farmer-cumrearer, reeler twister, weaver, hand spinners of silk waste, traders etc. It is the only one cash crop in agriculture sector that gives returns within 30 days.

This industry provides employment nearly to three five million people in our country. Sericulture is cultivated in Karnataka, Bengal, Tamil Nadu, Andhra Pradesh, Jammu & Kashmir, Gujarat, Kerala, Maharastra, Uttar Pradesh, Rajasthan, Bihar, Orissa etc.

Though India is the second largest silk producer in the World after China, it accounts for just 5% of the global silk market, since the bulk of Indian silk thread and silk cloth are consumed domestically. Germany is the largest consumer of Indian silk. The sericulture industry is landbased as silk worm rearing involves over 700,000 farm families and is concentrated in the three Southern states of Karnataka, Tamilnadu and Andhra Pradesh.

Silk, the queen of the fabrics still commands passion of consumer right from 2200 BC to till today, nationally and internationally. The export potential of Indian Sericulture Industry is evident from the fact that the annual export is Rs.2879.56 crores during the year 2004-05. The Central Silk Board, Ministry of Textiles, Govt. of India has been acting as a facilitator for planning, development and monitoring of sericulture industry between the States and Central Govt. The subject of Research is the exclusivity of the Central Silk Board and its sub-ordinate Research and Training Institutes.

Mulberry segment

Its main weakness is related to a poor database, diverse range of practices leading to a divergence in productivity and quality. Generally, there is weak accent on quality consistency in production, poor transfer of technology to the decentralised sector both due to poor technology absorption and poor/inadequate follow up on laboratory findings; poor market linkages barring in Karnataka, a thriving unfair trade in the post-yarn sector, low-end technology use and reluctance to costlier technologies due to fears that there might not be corresponding improvement in price realisations. Other weaknesses are inadequate emphasis on quality in the commercial seed sector, neglect of marketing linkages and the need for a basic perspective for development of the sector which clearly defined relative roles for the central and state agencies under the federal set-up.

Among non-mulberry silks, tassar is mostly produced by tribals by rearing silkworms on forest plants. India is the largest producer of tassar silk after China and is the only producer of golden muga silk. Also, India is a major producer of eri silk.

Unlike mulberry silk production, non-mulberry silk production is unsteady and fluctuates from year to year. The central silk board has not given enough attention to their R&D and extension activities in the area of non-mulberry sericulture in spite of its potential to directly help the poor. Presently, muga and eri silks are produced mostly for self-consumption. But with their uniqueness to India, they have great potential for value-added exports.

The government must give to these varieties of silk the importance that is due to them and facilitate focussed R&D, targeted extension and innovative product development for value-added exports.

Tassar

It has been noted that the following are the areas of weaknesses in production of tassar and they require to be set right

— Rearing is done outdoor on trees; natural food plants are dispersed over large areas. Thus, comprehensive extension support would entail a large number of extension agents to cater to the farmers beyond their resources.

— Also weavers are normally reelers and are not exploited by traders.

— Oak tassar culture has not yet been properly adopted, as people are new to this culture and economics are yet to be established.

— Also lack of disease monitoring and of control measures is noticed.

Eri

Eri silk which is largely produced in some eastern parts of the country has specific thermal proprieties. It can also be blended with wool, other silks, cotton ramie, jute or synthetic fibres. Areas of weakness in eri silk include:-

— Lack of systematic supply of adequate quantity of foilage.

— Lack of scientific method to check diseases

— Poor management during rearing

— Non-availability of separate rearing house and

— Absence of any well-organised marketing system

Muga

This golden yellow silk is quite unique to Assam and neighbouring areas of Nagaland and Meghalaya. It has spread to West Bengal and Andhra Pradesh as well. While basic seed production is more or less organised, commercial seed production is to be organised systematically.

3

Sericulture

Sericulture, or silk production, from the moth, *Bombyx mori* (L.), has a long and colorful history unknown to most people. Although there are several commercial species of silkworms, B. mori is the most widely used and intensively studied, and techniques for its rearing are the most improved. This insect is the sole living species in its family, Bombycidae, and has been domesticated for so long that it probably no longer survives in the wild.

According to Chinese records, the discovery of silk production from B. mori occurred about 2,700 B.C. Chinese legend states that the great prince, Hoang-ti, directed his wife, Si-ling-chi, to examine the silkworm and test the practicability of using the thread. Thereafter, Si-ling-chi discovered not only the means of raising silkworms, but also the manner of reeling the silk, and of employing it to make garments. Is-lingo-chi was later deified for her work and honored with the name Seine-Than, or "The Goddess of Silk Worms". Sericulture during the following centuries spread through China and silk became a precious commodity highly sought by other countries.

In 139 B.C., the world's longest highway was opened, and stretched from Eastern China to the Mediterranean sea. In addition to tangible commodities such as gold and jade, new ideas and religions also passed along this road. This road was the historically famous "Silk Road," named after its most important commodity. By the middle of the first century A.D., writers in Rome were complaining about the sumptuous silk garments that rendered women naked in the streets. But the Chinese had guarded the secrets of sericulture so closely the early Romans never learned it, and Virgil thought the thread was derived from combing the fuzz off leaves.

In spite of their secrecy, however, the Chinese were destined to lose their monopoly on silk production. Sericulture reached Japan through Korea, but not before the early part of the third century A.D. Shortly after 300 A.D., sericulture traveled Westward and the cultivation of the silkworm was established in India. According to tradition, the egg of the insect and the seed of the mulberry tree were carried to India concealed in the headdress of a Chinese princess. The emperor Justinian gained the secrets of sericulture for the Roman Empire in 522 A.D., with the smuggling of the silk worm eggs form China by Persian monks.

With China's monopoly on sericulture broken, silk importations from China became smaller and smaller. In 877 A.D., the rebel chief Biachu captured Canfu, the center of foreign silk trade, put to death all its inhabitants, destroyed all of the mulberry trees and silkworms of the region, and levied heavy and cruel taxes on all foreign trade. These actions stopped foreign commerce in China for more than 60 years. However, by this time, silk production was so well established in western Asia and eastern Europe that this wholesale destruction hardly effected the price of silk in the rest of the world.

During the 18th and 19th centuries, Europeans also produced several major advancements in silk production. England by the 18th century led Europe in silk manufacturing because of English innovations in the textile industry. These innovations included improved silk-weaving looms, power looms and roller printing. In 1801, A Frenchman named Joseph Jacquard exhibited his new machine foe figured-silk weaving and gradually spread through the industry. The great French scientist, Louis Pasteur, rescued the silk industry in 1870 by showing that the then epidemic Pebrine disease of silk-worms could be controlled by prevention through simple microscopic examination of adult moths. These advances set the trend for a more mechanized and scientific approach to silk production than existed previously.

Sericulture has also been attempted in the United States, but these endeavors have been sporadic and largely unsuccessful. Sericulture was carried on to some extent by the early colonists of Virginia, South Carolina, and Georgia, and was introduced into New England about 1660. In 1831, a manual on sericulture was published by J.H. Cobb, copies of which were purchased by the Congress of the United States for distribution by members. Following publication of this book, there was a determined effort to establish silk culture on a firm basis in the United States. This interest in silk culture soon led to what was known as the "Mormus multicaulis craze."

Anticipating a most profitable investment, if not speedy riches, thousands of individuals purchased mulberry plants of the M. multicaulis species and planted large areas of valuable land. The investments far exceeded possible returns, and heavy frosts destroyed plantations of trees. In the course of a few years, many failures and great disappointments caused so complete a revulsion of feeling that silk culture was practically abandoned all through the States. However, because confederate cotton was unavailable during and shortly after the Civil War, the Union States were forced to seek a new source of fiber.

Thus in 1869, Professor L. Trouvelot, an American naturalist, brought eggs of the gypsy moth, Lymantria dispar (L.), from France to Massachusetts. Trouvelot had hoped to produce a commercial source of silk by developing a hardy race of silk-producing insects, crossing the gypsy moth with the silkworm moth, in order to control wilt disease (or flancheria) then causing severe problems in some silkworm industries. However, during the course of his experiments, some of the eggs were lost and some of the caterpillars escaped from his home. Although this accident was made public at the time it did not receive much attention even though the gypsy moth was immediately recognized as a pest. Since its introduction into the Boston area over a century ago, the gypsy moth has greatly expanded its range and become one of North America's most serious forest pests, defoliating large areas of canopy every year.

In spite of these earlier failures at sericulture in the United States, several more attempts at sericulture were made in California from the 1860's through the early 1900's. California sericulturists even advocated the commercial rearing of the native ceanothus silk moth, Hyalophara euryalus (Boisduval), as a possible source of silk until Felix Gillet in 1879 showed that the cocoons could not be reeled satisfactorily. Although some silk was produced in California during this time, most sericulture attempts failed and sericulture never became permanently established in the state.

Silk production today is a blend of ancient techniques and modern innovations. The first stage of silk production is hatching the silkworm eggs, which have been previously examined and shown to be free from disease. Larvae are then fed cut-up mulberry leaves and after the fourth molt climb a twig placed near them and spin their silken cocoons. The silk is a continuous-filament fiber consisting of fibroin protein secreted from two salivary glands in the head of each larvae, and a gum called sericin, which cements the two filaments together.

Pupae within cocoons are killed by steam or fumigation to prevent adult emergence, which would cut and tangle the silk filaments. Cocoons are latter softened in hot water to remove the sericin, thus freeing silk filaments for reeling. Single filaments are drawn from cocoons in water bowls and combined to form yarn. This yarn is drawn under tension through several guides and eventually wound onto reels. The yarn is dried, packed according to quality, and is now raw silk ready for marketing.

World silk production has approximately doubled during the last 30 years in spite of man-made fibers replacing silk for some uses. China and Japan during this period have been the two main producers, together manufacturing more than 50% of the world production each year. China during the late 1970's drastically increased its silk production and became the world's leading producer of silk. The 1970's were a period of tumultuous political and social upheaval in China, resulting in various economic reforms. Undoubtedly, these reforms are partially responsible for China's increased silk production. Thus the country that first developed sericulture approximately 4,700 years ago has again become the world's main producer of silk.

Chinese Contribution

Silk is a great invention of the Chinese people and one of its significant contributions to world civilization. Such and such legends about the origin of silk are tinted with romantic mythical tones. China's silk culture also affords inexhaustible fascination.

As early as 5,000 years ago the Chinese have successfully domesticated the wild silkworm by using the silk it spins to make cloths. Judging from the distribution of silk relics in China silk production spread to most parts of China long time ago and there arose some regions renowned for silk products, with the middle and lower reaches of the Yellow River being the center of production.

North China was inflicted with wars after the third century while economy in the south was developing steadily by comparison, which moved silk production center to the south. The middle and lower reaches of the Yangtze River, particularly the south of the lower reaches, had rapid development after the An-Shi Rebellion, and a situation of tripartite balance basically formed among the lower reaches of the Yellow River, Sichuan and the lower reaches of the Yangtze River in late Tang Dynasty. When the Song authorities moved government to the south silk production was even more flourishing there, which gave rise to many towns specialized in silk production and silk markets dependent on them in the south of the Yangtze River lower reaches. Since Yuan Dynasty, cotton plantation began to spread nationwide and silk production was further concentrated in the Yangtze River lower reaches. A situation formed where the south made a great deal of silk and the north consumed it in huge quantities.

In the perpetual river of history silk has pervaded every aspect of Chinese society through the ages and has significant influence on people's life and production. It has gradually formed a unique silk culture that became the cream of the Chinese people's cultural treasures. The evolution of silk in ancient China has enriched people's life not only in terms of clothing but also many artistic expressions such as literature, poetry, painting, sculpture, dancing and folk customs etc.

Countless ties with silk can always be found in these areas. The world-renowned Silk Road was a cultural and commercial tie between the east and the west. Linking the remote Greek and Persian cultures it promoted the commercial and trade exchanges and enhanced the friendship among various countries. China thus won the laudatory title of Seres (silk country).

Silk clothes and articles used to be necessities of the luxurious life of aristocrats, which led to refined quality of weavings by state-run firms catering to the up market. The development of production demanded advance of techniques, which in turn put China in the leading position in the world in terms of silk manufacturing techniques for a considerable period. The constant innovation of mulberry plantation techniques, weaving techniques and silk weaving arts as well as the inventions by numerous craftsmen consistently pushed the advance of silk-related techniques and artistic level.

We can get a glimpse of the glorious development through cultural relics and articles handed down from ancient times. In Han Dynasty China witnessed its first peak of silk industry. There was great variety of silk products weaving techniques. The level of silk development can be sufficiently testified by the cultural relics excavated at Mawangdui in Hunan. It was during that time that the Silk Road was officially opened.

Silk was already famed worldwide in Tang Dynasty and it was the clothe for the whole world. The masterly products, exquisite grain and flowery colors all combined to make the Tang Dynasty silk the perfect integration of art and techniques. China's silk weaving and embroidery techniques made a historic leap in Song Dynasty. Kesi (silk tapestry with cut designs) and embroidery techniques were very much matured. It was a vogue to copy from such artworks as realistic paintings of landscape and flowers and birds, calligraphy etc. leaving exotic flowers in silk history. In Ming and Qing Dynasty silk developed toward a more refined and aristocratic direction, displaying a resplendent, sumptuous and magnificent vigor and driving the pattern design and weaving techniques to the superlative.

For 5,000 years China's silk has written a splendid chapter in history, reaching an astonishing level both in terms of art and techniques. The rich variety, fine quality, exquisite pattern and grain and masterly crafts all make people acclaim as the peak of perfection! For example, the formation of Kesi, Zhijin (picture-weaving in silk) and embroidery represents the high level China's silk has accomplished and the great wisdom of the Chinese people. These traditional weaving techniques are insurmountable even with today's high-tech means.

Silk

Silk is a natural protein fibre, some forms of which can be woven into textiles. The best-known type of silk is obtained from the cocoon of silkworm larvae reared in captivity (sericulture). The shimmering appearance for which silk is prized comes from the fibres' triangular prism-like structure which allows silk cloth to refract incoming light at different angles.

Silk fabric was first developed in ancient China, possibly as early as 6000 BC and definitely by 3000 BC. Legend gives credit to a Chinese Empress Xi Ling-Shi. Though first reserved for the Emperors of China, its use spread gradually through Chinese culture both geographically and socially. From there, silken garments began to reach regions throughout Asia. Silk rapidly became a popular luxury fabric in the many areas accessible to Chinese merchants, because of its texture and lustre. Because of the high demand for the fabric, silk was one of the staples of international trade prior to industrialization.

The Emperors of China strove to keep the knowledge of sericulture secret from other nations, in order to maintain the Chinese monopoly on its production. This effort had mixed success. Sericulture reached Korea around 200 BC with Chinese settlers, about the first half of the 1st century AD in Khotan, and by 300 AD the practice had been established in India. Although the Roman Empire knew of and traded in silk, the secret was only to reach Europe around AD 550, via the Byzantine Empire.

Legend has it that the monks working for the emperor Justinian were the first to bring silkworm eggs to Constantinople in hollow canes. The Byzantines were equally secretive, and for many centuries the weaving and trading of silk fabric was a strict imperial monopoly; all top-quality looms and weavers were located inside the Palace complex in Constantinople and the cloth produced was used in imperial robes or in diplomacy, as gifts to foreign dignitaries. The remainder was sold at exorbitant prices.

Silk Trade

Perhaps the first evidence of the silk trade is that of an

Egyptian mummy of 1070 BC. In subsequent centuries, the silk trade reached as far as the Indian subcontinent, the Middle East, Europe, and North Africa. This trade was so extensive that the major set of trade routes between Europe and Asia has become known as the Silk Road.

"Wild silks" are produced by a number of undomesticated silkworms. Aside from differences in colours and textures, they all differ in one major respect from the domesticated varieties. The cocoons that are gathered in the wild have usually already been damaged by the emerging moth before the cocoons are gathered, and thus the single thread which makes up the cocoon has been torn into shorter lengths.

A variety of wild silks have been known and used in China, India, Vietnam, and Europe from early times, although the scale of production has always been far smaller than that of cultivated silks. Wild silks are produced by caterpillars other than the mulberry silkworm (Bombyx mori). The term "wild" implies that these silkworms are not capable of being domesticated and artificially cultivated like the mulberry worms.

Commercially reared silkworm pupae are killed before the adult moths emerge by dipping them in boiling water or piercing them with a needle, thus allowing the whole cocoon to be unravelled as one continuous thread. This allows a much stronger cloth to be woven from the silk. Wild silks also tend to be more difficult to dye than silk from the cultivated silkworm. There is some evidence that small quantities of wild silk were already being produced in the Mediterranean area and the Middle East by the time the superior, and stronger, cultivated silk from China began to be imported.

Many different types of silk are produced by a huge variety of different types of insect (other than moth caterpillars), yet none of these have been exploited for commercial purposes, though there is basic research into the structures of such silks, as there is some variation at the molecular level. It is most commonly produced by larvae, and thus largely limited to insects with complete metamorphosis, but in some cases it is produced by adult insects such as webspinners. Silk production is especially common in the Hymenoptera (bees, wasps, and ants), and is sometimes used in nest construction. Other types of arthropod produce silk, most notably various arachnids such as spiders.

Europe

Venetian merchants traded extensively in silk and encouraged silk growers to settle in Italy. By the 13th century, Italian silk was a significant source of trade. Since that period, the silk worked in the province of Como has been the most valuable silk in the world. Italian silk was so popular in Europe that Francis I of France invited Italian silkmakers to France to create a French silk industry, especially in Lyon. Mass emigration (especially of Huguenots) during periods of religious dispute had seriously damaged French industry and introduced these various textile industries, including silk,to other countries.

James I attempted to establish silk production in England, purchasing and planting 100,000 mulberry trees, some on land adjacent to Hampton Court Palace, but they were of a species unsuited to the silk worms and the attempt failed. Production started elsewhere later. Silk was also used in Medieval Europe for the Kings, Queens, Lords and Lady's.

India

Silk, known as 'pattu', is widely produced in India today. Historically silk was used by the upper classes, while cotton was used by the poorer classes. Today silk is mainly produced in Kanchipuram, Dharmavaram, etc., in South India and Banaras in the North. The silk from Kanchi is particularly well-known for its classic designs and enduring quality. The silk is traditionally hand-woven and hand-dyed and usually also has silver threads woven into the cloth. Most of this silk is used to make saris. Garments made from silk form an integral part of Indian weddings and other celebrations.The saris usually are very expensive and vibrant in colors.

North America

James I of England introduced silk-growing to the American colonies around 1619, ostensibly to discourage tobacco planting. Only the Shakers in Kentucky adopted the practice. In the 1800s a new attempt at a silk industry began with European-born workers in Paterson, New Jersey, and the city became a US silk centre, although Japanese imports were still more important.

World War II interrupted the silk trade from Japan. Silk prices increased dramatically, and US industry begun to look for substitutes, which led to the use of synthetics like nylon. Synthetic silks have also been made from lyocell, a type of cellulose fibre, and are often difficult to distinguish from real silk.

Islam

In Islamic teachings, Muslim men are forbidden to wear silk. While the command is given without justification, many religious jurists believe the reasoning behind the prohibition lies in avoiding clothing for men that can be feminine or extravagant and luxurious. Furthermore, Islam also prohibits the slaughter of animals for "sport" or "luxuries", hence the Islamic prohibition against the use of silk (a single pound of which requires the killing of 3000 silkworms, according to PETA.). These traditions are not, however, universal in the Muslim community. For example, Muslim men who wear Western business suits generally wear silk ties.

Muslims/Moors brought silk with them to Spain during their conquest of the Iberian Peninsula. After the reconquista, the Muslims and the Jews were expelled in 1492; then Columbus sailed East—toward Asia, he hoped, looking for gold, silver, silk, and spices. He brought wool to trade. He found little gold and silver—and no silk—because he had not really reached China.

Uses

In addition to clothing manufacture and other handicrafts, silk is also used for items like parachutes, bicycle tires, comforter filling and artillery gunpowder bags. Early bulletproof vests were also made from silk in the era of blackpowder weapons until roughly World War I. Silk undergoes a special manufacturing process to make it suitable for use as non-absorbable surgical sutures. Chinese doctors have also used it to make prosthetic arteries. Silk cloth is also used as a material to write on.

Mongols used silk as part of the under-armor garments. Silk is so tough that it was actually used as very light armor, although its special use (the big secret) was to stop arrow penetration into the body. The silk would stop an arrow from penetrating far enough into the body to be lethal; and the arrow could then be pulled out of the wound by tugging on the unbroken silk. The added advantage to this is that there would be no contact between the arrowhead and the interior of the body; thus it reduces the incidence of infected wounds.

Grasserie—Disease of Silkworms

Grasserie is caused by a virus called Borrelina. It is infectious through contact with the effluent from the sick worms, and it can be spread through the skin of the silkworms and through ingestion of contaminated food. The disease is also called nuclear polyhedrosis virus, and is listed as a disease that attacks several species of wild North American insects.

The first sign is that the worm looks a lot more plump than it should. When it's stretched out, it's kind of hard to tell - but the segments don't look right. When the worm contracts, the over-fullness of the segments is distinct. They look like little sickly Michelin Men. They also become very fragile - the skin breaks easily, and oozes infectious milky liquid. It reminds me a lot of proofing yeast in water.

The worms start to get dark. Some of them will get what look like hernias, where the green mulberry inside pushes out toward one side. They get listless, and stop eating. They eventually stop moving, although it takes them a while to die. Once dead, within just a few hours they are black and putrefying. You have to be very careful picking them up, or they bust open and ooze guts. It's really gross.

Hygeine is critical in dealing with disease in silkworms. The disease can be managed with several methods:

— immediately remove and discard any worms that start to show symptoms, before they can ooze on the leaves or other worms.

— wash the trays, and then disinfect them with bleach water; wash the netting between tray cleanings with bleach.

— net the worms up daily, or even twice daily - the sick ones don't climb upward, so it separates the well from the ill.

— wash hands and spritz them with specially denatured high-proof alcohol any time you have to handle infected worms or trays.

4

Bombyx Mori

Bombyx mori, is a member of a small family of about 300 moth species in the Order Lepidoptera. *Bombyx mori* is not native to North America, but rearing here is prevalent. Adults in this family have heavy, rounded, furry bodies and cannot feed because of undeveloped mouth parts. The forewing has a hooked tip, characteristic in this family, however it is flightless. Wings and body are usually white, but may vary to shades of light brown. Wingspan is 1.5 to 2.5 in. (4-6 cm).

The larva of *Bombyx mori* is an elongated caterpillar commonly called a silkworm. Larvae are monophagous and feed only on mulberry plants. The larva will molt four times before spinning a silk cocoon of one continuous fiber for pupation. Silk cocoons from this species are the commercial source of silk. Bred in captivity for thousands of years on trays of white mulberry (*Morus alba*), *Bombyx mori* is fully domesticated and cannot survive without the assistance of man. Although no wild colonies remain, scientists believe the origin of *Bombyx mori* is likely in Asia.

As a valuable commodity in the textile market, rearing is common in Europe, Asia and South America. Useful in education and research, rearing the species is practical in most parts of the U.S. where a supply of fresh mulberry is available. The process of rearing *Bombyx mori* is simple and the same principles apply for any application or any size project; a controlled environment and a supply of fresh mulberry leaves.

Rearing silkworms is an indoor project for several reasons. The temperature range necessary for the insect (average room temperature) is easily regulated within an indoor space. With no defence mechanisms, *Bombyx mori* is also protected from predators when reared inside. With good hygiene an important component of rearing, a clean rearing area is easy to accomplish in an indoor environment.

Mulberry (*Morus*) leaves are the food of *Bombyx mori.* Since mulberry plants are deciduous, start a rearing project in the spring or summer when an adequate supply of fresh leaves are available. Rearing silkworms on artificial diet formulas of leaf powder and protein are not recommended for most of the commercial silkworm strains currently available.

Kingdom- **Animalia**
- Phylum- **Arthropoda**
  - Class- **Insecta**
    - Order- **Lepidoptera**
      - Family- **Bombycidae**
        - Genus- **Bombyx**
          - Species- **mori**

Figure 1. Classification of Bombyx mori L.

The most economical and simple way to rear silkworms is in a tray or shallow box. Immediately after hatching, provide larvae with clean, fresh leaves cut to dime-size pieces. As the larvae grow, provide larger pieces or full leaves. Feed the larvae a few times per day or as frequently as needed to keep leaves available for consumption. About 27 days after hatching, silkworms begin spinning cocoons. Providing a support mechanism like an egg carton, partitioned box or crumpled paper facilitates efficient construction of the cocoons. Cocoon completion takes about 3 days.

Cocoons used for reeling must be baked in an oven to prevent moths from piercing their cocoons and breaking the continuous fiber. If left to develop naturally, the adult silk moth emerges from the cocoon about two weeks after completion. Soon after eclosion, moths mate and females lay eggs. The life cycle of *Bombyx mori* demonstrates the most advanced form of metamorphosis. Termed holometabolous, the serial progressions of four distinct stages of development complete one generation of the species; ova, larva, pupa and imago.

The number of generations per year or season depends on the voltinism of the silkworm strain and variables including temperature. Voltinism is when some members of a species enter a hibernation-like period of diapause, while others do not. Under natural conditions, silkworm strains producing one generation per year are univoltine. After removal from cold storage, ova in diapause begin final development until hatching, about two weeks later. About six to eight weeks after the start of ova incubation, one generation of *Bombyx mori* is complete.

Stage 1 - OVA: Incubation 10-14 Days

The egg of *Bombyx mori* is a very small and hard structure; about the size of a pin head and resembling a poppy seed. The egg shell provides a protective covering for embryonic development. When first laid, an egg is light yellow. Fertile ova darkens to a blue-gray within a few days.

Stage 2 - Larva: 27 Days (5 instars)

The larva is the vegetative stage where growth takes place. The larva of *Bombyx mori,* commonly called a silkworm, is host specific to mulberry. During growth, the larva will molt 4 times. The period between successive molts is called an instar.

State 3 - Pupa: 14 Days

The silk cocoon serves as protection for the pupa. Cocoons are shades of white, cream and yellow depending on silkworm genetics. After a final molt inside the cocoon, the larva develops into the brown, chitin covered structure called the pupa. Metamorphic changes of the pupa result in an emerging moth.

Stage 4 - Imago : 5 - 7 Days

The adult stage completes the life cycle of *Bombyx mori.* It is the reproductive stage where adults mate and females lay eggs. Moths are flightless and lack functional mouth parts, so are unable to consume nutrition. Success in rearing *Bombyx mori* begins with a simple plan. Without regard to the size or purpose of a rearing project, start with a written plan. Develop the project objectives, then expand the project design with details that will accomplish the goals.

A significant aspect of rearing any organism is to consider the living material that requires attention. Within your general plan, include such specifics as a start date, a plan for leaf harvest, feeding responsibilities, schedules, etc. Finally, set up a log or journal to record data that may be useful for future rearing projects.

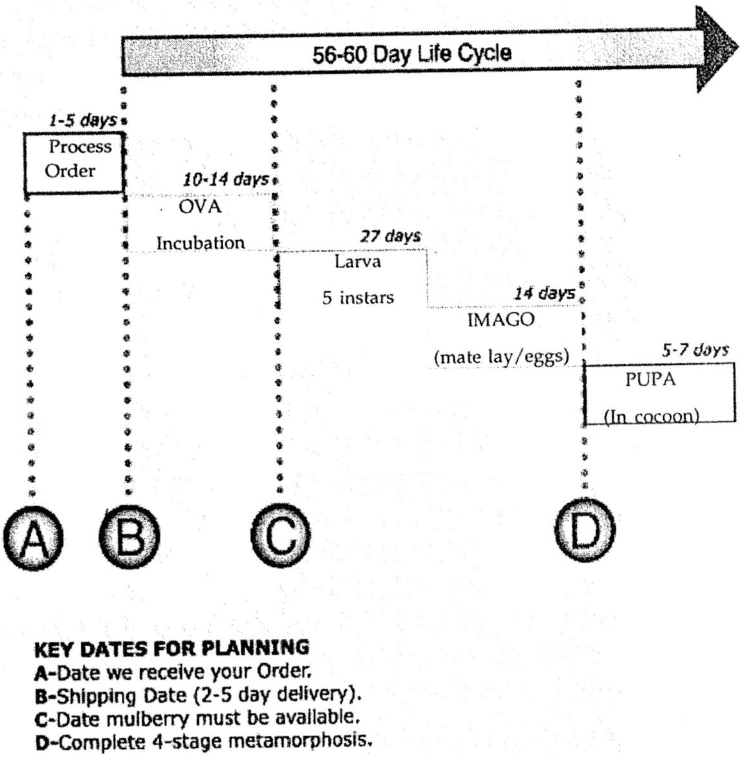

Figure 2. Silkworm Rearing Planner

Artificial Insect Diet

As dietary replacements or supplements, artificial insect diet formulas are mixtures of various ingredients to sustain the feeding stages of a life cycle. Best utilised in a laboratory setting, these synthetic foods attempt to provide insects with nutritional requirements for growth and development. In silkworm rearing, scientists may rely on artificial diets within the confines of small-scale, controlled experiments where life cycle completion is not a prerequisite.

In some commercial silk production, sericulturists may use artificial diet formulas designed for strains of silkworms that are genetically suited for feeding on the synthetic diets. Since the nutritional requirements change as larvae develop, silkworms are routinely provided with different ages of mulberry leaves that are best for their stage of growth. It is generally accepted that different formulas of artificial diet are required for each stage of larvae development as well, especially if optimum silk cocoon spinning is desired.

Single diet formulas for unsuitable silkworm strains may not provide adequate nutrition for the changing dietary needs as larvae grow and mature. As a result of poor nutrition, varied forms of mutation and retardation in development regularly occur. Although scientists do not have a complete understanding of the mechanism, an example of a recessive mutation is one in which bitter taste receptors are stimulated on the palpi of larvae, causing an aversion to feeding. Poor feeding habits result in slow growth or mortality.

In continued study of this subject with testing and assessment of published formulas and various commercially available artificial silkworm diets, the following general abnormalities were observed in rearing examinations of several commonly available strains of *Bombyx mori*, by the Sericulum Company:

— An irregular growth rate and slow instar progression.

— Larval mortality rates were regularly high and unacceptable.

— Cocoons were usually small and thin walled.

— Lowered percentage of ecloding adults.

— Adult deformities including minute wings and missing antennae.

— Reduced pairing/mating of adults.

— Small numbers of laid ova and irregular oviposition.

— Reduced ova fertility, viability.

Beyond the complex genetic traits that may inhibit normal development, other limitations of many artificial silkworm diets include a restricted shelf life (many are sold repackaged with no original manufacturers dating) and a lack of complete content labelling with ingredient percentages (most contain antibiotics). Additionally, the good quality, multi-stage diets available from Japan are relatively expensive and rarely sold in weights under 20lb.

For the end user, the use of artificial diets also demand an increase in the labour required to regularly prepare the formulas for freshness and to maintain hygienic conditions within rearing containers. A consideration for those demonstrating the life cycle of silkworms is one of the visual presentation associated with the use of artificial diets. Diet formulas are typically mixed with boiling water to form viscous consistencies resembling cookie dough. When utilised in a classroom, for example, these preparations often collect larvae frass with odorous and unpleasant effects.

Most important, the use of artificial silkworm diets fail to demonstrate the centuries-old practice of feeding silkworms on trays of mulberry leaves. As a result of the scientific challenges, the risk of non-completion of a normal, four stage life cycle for unsuitable silkworm strains and the poor esthetics that come with artificial diet use, the Sericulum Company does not offer them for sale.

Mulberry

The Mulberry family of plants, *Moraceae*, comprises more than 50 genera and 1,000 species. The Mulberry genus, *Morus*, is a group of about a dozen species, although so variable that descriptions exist for many times that amount. Mulberries are native to the temperate and subtropical regions, most in Asia. Two are native to North America; *Morus rubra* (Red Mulberry) and the lessor known *Morus microphylla Buck* (Texas Mulberry). Two species native to Asia, Morus nigra (Black Mulberry) and *Morus alba* (White or Silkworm Mulberry) have become naturalised in North America after introduction in the 1600's.

Cultivated for centuries, mulberries are grown for their edible fruit, ornamental use in landscapes and as the host plant for *Bombyx mori*. White Mulberry (*M. alba*) is the most important for horticultural use and provides the best food source for silkworms. Leaves of the deciduous *Morus alba* plants are varied in form, size and shape. Blackberry-like fruits are sweet but tart. Home gardeners grow fruiting mulberry for use in jams, jellies and for making wine. Fruitless varieties of *Morus alba* are desirable for landscape use. Fast growing, relatively easy to grow, tolerance of heat and most soil types, make mulberry a popular choice in horticulture across the country.

Silkworm Needs

Natural Food

Rearing habitat

— Petri dishes or pint size containers

— An aquarium or wide mouthed jar with a lid

— For larva: A food plant vase made from film can or 1/2 pint container with "x" cut in lid

— Pruning shears or heavy scissors are helpful to gather plant material

— Laundry bleach (sodium hypochlorite)

The natural food plant of the silkworms is the mulberry tree (*Morus* sp.). Make sure you have a dependable source of pesticide free mulberry leaves before beginning this project. Before feeding the leaves to the larvae, soak the leaves for three minutes in a cleaning solution (three tablespoons of laundry bleach and a drop of dish washing detergent in a gallon of water). Thoroughly rinse the leaves in running water and shake off excess water. Store the leaves in a plastic bag in the refrigerator. The first through third instar larvae require young, tender leaves of mulberry.

Offer a leaf in a petri dish and change the leaf daily. After a week you will need to supply more than one leaf. A small piece of damp paper wrapped around the end of the leaf will prolong the freshness. The fourth and fifth instar larvae will need more room and can be set up in the gallon jar with a container to hold mulberry stems and leaves. These older larvae can eat older mulberry leaves. Recut stem ends and quickly insert into water when changing the food plant.

Water

Larvae do not need water if the natural food plant is used because they can obtain all the water they need from the plant. If you cannot obtain the natural food plant, you must use the artificial diet.

Artificial Food

Rearing Habitat

— Disinfected glass petri dishes

— Filter paper or paper towelling

— Spray mister

— Circular sheet of paraffin paper which fits inside the petri dish

— Enough artificial diet for each larva instar

— Alcohol to reduce fungal growth on artificial food.

First instar larva: Dip one bar of diet in water and remove when the water is absorbed. Remove excess water with paper towel. Place a moist piece of filter paper or paper towel in the bottom of the petri dish and cover with a sheet of paraffin paper. Place the moistened diet on the paraffin paper and transfer the larvae. Cover the diet and larva with a sheet of paraffin paper. Check the diet daily and add drops of water to keep it moist.

Second and Third instar larva: After four days, the larva will stop eating and molt. After molting, clean and disinfect the dish or use a new sterile dish. Use the same setup without the moist filter paper. The third instar larva will need at least two petri dishes and will eat at least two bars of diet.

Fourth and Fifth instar larva: Make sure the larva have enough room and provide additional petri dishes as needed. The top sheet of paraffin paper is no longer needed. Continue to check the diet and keep it moist. If mold appears, remove the larvae and add one or two drops of disinfectant alcohol. Allow the alcohol to evaporate and reassemble setup.

Water

The artificial diet must be moistened with water before the larvae can eat it.

Cleaning

— Frass (droppings) needs to be removed frequently to discourage mold. Gently remove the water container and dump out the frass. Every time you change the food plant, rinse out the container and thoroughly dry it. Periodically wash the water container for the food plant.

— Replace the filter paper and paraffin paper every time you give the larvae new artificial diet. Sterile petri dishes are strongly recommended to reduce fungal growth.

Newly hatched larvae can be transferred by using a small paint brush. Older larvae can be gently picked up with your hands. When the fifth instar larva begins to shrink (after sixth to eighth day), prepare a cocoon nest. Using cellophane paper or paper towels, make a roll of paper by twisting one end and placing two or more larvae in it. Twist the other end closed. Store the rolls in the dark at 25°C (77°F). It takes the larva three days to spin the cocoon and then two to three days to molt into a pupa. By cutting the end of the cocoon, a pupa can be gently removed from the cocoon for examination and returned to the cocoon after observation. Although the adults do not fly, the moths have delicate wings that can be damaged by handling.

If you ordered silkworm eggs, the eggs should be placed in a petri dish. After seven days, the eggs will begin to hatch and continue for one to two weeks. Once the first eggs have hatched, check the dish every day for new hatchlings. A day or two before hatching, the egg will darken around the edge. Transfer newly hatched larvae with a fine paint brush to artificial culture or food plant. If adults copulate in captivity, the female will lay eggs within 24 hours. These eggs will not hatch for some time and can be stored.

If you commit to feeding the caterpillar mulberry leaves, make sure you can obtain plenty of leaves. Once the caterpillars have eaten leaves they will not eat the artificial diet. Try to maintain the temperature between 68°F to 86°F especially during the first three instars. If you have access to an incubator, keep it at 84°F (29°C). The later instars can be reared with or without an incubator. Do not place the rearing container where direct sunlight will hit the container. The larvae can easily overheat and die.

Silk Production

The cocoon is made of a single continuous thread of raw silk from 300 to 900 meters (1000 to 3000 feet) long. The fibers are very fine and lustrous, about 10 nanometers (1/2500th of an inch) in diameter. About 2,000 to 3,000 cocoons are required to make a pound of silk. Based on 1 kilometer (2/3 of mile) per cocoon, ten unraveled cocoons could theoretically extend vertically to the height of Mt Everest. It is estimated that at least 70 million pounds of raw silk are produced each year, requiring nearly 10 billion pounds of mulberry leaves.

According to E. L. Palmer, one pound of silk represents about 1,000 miles of filament. The annual world production represents 70 billion miles of silk filament, a distance well over 300 round trips to the sun. If the animal is allowed to survive after spinning its cocoon, it will make a hole in the cocoon when it exits as a moth. This would cut short the threads and ruin the silk. Instead, silkworm cocoons are thrown into boiling water, which kills the silkworms and also makes the cocoons easier to unravel. Often, the silkworm itself is eaten.

The adult phase (the moth) cannot fly. The silkworm-moths have a wingspan of 2 inches and a white hairy body. Females are about twice to three times as large as the males, but similarly coloured. Adults in the Bombycidae have reduced mouthparts and do not feed.

Scientific Uses

Due to its large size and ease of culture, the silkworm has long been a model insect in the study Lepidopteran and arthropod biology. Fundamental findings on pheromones, hormones, brain structures and physiology were made with the silkworm. To characterize the first known pheromone, bombykol, extracts were needed from 500,000 individuals because only very small quantities are produced. Currently, research is focusing on genetics of silkworms and genetic engineering.

Many hundreds of strains are maintained, and over 400 Mendelian mutants have been described. One useful mutant for the silk industry confers the ability to feed on food besides mulberry leaves, including an artificial diet. The genome is being sequenced, and many projects have worked on genetic engineering of silkworms to produce desirable proteins in the place of silk. Such proteins include human drugs.

In China, there is a legend that the discovery of the silkworm's silk was by an ancient empress called Xi Ling-Shi. She was drinking tea under a tree when a cocoon fell into her tea. She picked it out and as it started to wrap around her finger, she slowly felt a warm sensation. When the silk ran out, she saw a small cocoon. In an instant, she realized that this cocoon was the source of the silk. She taught this to the people and it became widespread.

There are many more legends about the silkworm. The Chinese jealously guarded their knowledge of silk. It is said that a Chinese princess smuggled eggs to Japan, hidden in her hair. The Japanese thus began their love affair with silk. It takes 2100 silk worms to make a single kimono.

Silkworm Diseases

One of the silkworm's deadliest enemies are ants, who devour the silkworms. But it is easy to confuse young, tiny silkworms with ants, as they look somewhat alike. Silkworms suffer from infectious diseases caused by protozoans, fungi, viruses, and bacteria. The French microbiologist Louis Pasteur investigated several silkworm diseases, which threatened the European silk production at that time.

Medical Uses

Silkworm is the source of the traditional Chinese medicine "Bombyx batryticatus" or "stiff silkworm". It is the dried body of the 4–5th stage larva which has died of the white muscardine disease caused by the infection of the fungus Beauveria bassiana. Its uses are to dispel wind, dissolve phlegm and relieve spasm.

Cuisine

In Korea, silkworm pupae, boiled and seasoned, are a popular snack food known as beondaegi. In China, street vendors sell roasted silkworm pupae. Sometimes the pupae are stir-fried and eaten with salt and powdered spice.

5

Varieties in Silkworms

Daudolo says, a quantity of these silk-worms separately, the eggs of which may be, found in several parts of Lombardy. The worms and cocoons of this variety are two-fifths smaller than those of the common sort. His experiments demonstrated, drat these worms consume, to form a pound of cocoons, nearly as much leaves as the large species; and, although smaller when they have reached their full growth, they devour more fragments and shoots of leaves than the common sort.

Their cocoons are composed of fitter and mote beautiful silk than the common cocoon: they arc also better constructed, and to this is owing the greater quantity of silk which, at equal weights, is drawn from those cocoons, than from common cocoons. This variety, in the opinion of Darrdolo, should be infinitely snore cultivated than it has hitherto been : for

1. These silkworms require four days less of care than the common silk-worms.
2. They afford a saving of time, labour, and money.
3. They are not so long exposed to accidents—their life being shorter. Some imagine the species to be delicate, but they appeared to Dandolo to be strong

and vigorous. Six hundred cocoons weigh't pound and a half.

Casts of Silk-worms

Dandolo reared many of these silk-worms of a very large quality. The eggs came from Friuli: they produce larger worms and larger cocoons, yet they are not much larger nor much heavier than the eggs of the common species. The worms, at their full size, weigh nearly twice and a half as much as the common worm. The cocoons are, in the same proportion: 150 of the large sort weigh a pound and a half, while it requires 360 of the common cocoon to weigh as much. The silk is coarser and not so pure as that of the common kind; the worms are five or six days later in attaining their full growth, and in rising, than the common silk-worm.

Says Dandolo, "a large quantity of these, and found them, in all respects, equal to the common silk-worms of four casts. This species was introduced into France, about forty-five years since, from China, but was nut much cultivated until about seventeen years past. It is now highly prized by the manufacturers, as appears by the report of the exhibition of French Industry in the year 1819.

In Windham county, Connecticut, there is also a small pale: white worm, which eats but twenty days, and produces fine white silk, though in less quantity than either the common large pale white, or the dark coloured worm ;but it has the good duality of retaining its clear white colour, and dues nut turn yellow by washing, or by exposure to son and air. These worms produce also two crops. It is highly probable, that these white worms are of the same species as that last mentioned.

The dark drab coloured worms, which are very common in the United States, and called "black," live longer, anti make snore silk than the large write worms.

Silk-worms of Eight Crops

At the silk establishment of the British East India company at Jungepore, Bengal, Lord Valencia, besides the common annual silk-worm which gave but one crop, found two others; the one commonly reared, and supposed to be indigenous, is called Dacey, producing eight harvests. Another and worst, the China, or Madrassa, also yields eight tunes a year. This last may be the kind mentioned by Another Young, who says, he obtained a silk-worm from China, which he reared, nod, in twenty-five slays, had the cocoons in his basins, and, by the twenty-ninth or thirty-first days, a new progeny feeding in his trays." He justly remarks, that "they would he a mine to whoever would cultivate them." The American, who would introduce any of the best of these silk worms into the United States, would render an essential service to his country.

Worm eggs from Philadelphia, which were kept in an upper room, without a fire, and hatched early in April: they were fed the first week on lettuce, afterwards on the leaves of the white and native mulberry tree, and came to maturity within from twenty-lour to thirty days, and spun their cocoons before the 12th of May. On the last days of May, and first of June, the moths came out anti laid their eggs on paper, which was loosely, rolled up, and placed in an upper room. Within eight or ten days from the time the eggs were laid, they began to hatch, and before the 15th of June, nearly one-twentieth part had hatched; the worms were healthy, fed well, grew more rapidly, and came to maturity in a few days less time, than the first crop.

They wound their cocoons in the first days of July; on the last of that month, and first of August, the moths came forth, and laid their eggs, which were placed in the same room as the first eggs. In about the same length of time, they also began to hatch, and all the worms came out between the 12th and 18th of August. They cattle to maturity, and spun their cocoons, in somewhat less time than the first, viz. before the middle of September: the moths came forth, anti laid their eggs the first days of October."

Mosses Weiss and Youngman, of Bethlehem, Pennsylvania, raised two crops during the year 1825. The worms of the second crop appeared more vigorous and healthy, than those of the first. They also produced larger cocoons, the silk of which, was of a better quality." Two attempts, viz: in 1826 and 1827, to rear two crops of worms, by Messrs. Terhoeven, Philadelphia county, failed. The worms, from the eggs laid in the early part of the season, hatched, but they soon died.

Caterpillars Producing Silk

The larva of the silk moth, is not the only insect by which silk is spun. Other insects produce the same substance, which, in point of strength, far exceeds that of the common silk-worm. These insects are natives of India, and bear the names of Bughy or Tnsseh, and Arrindy silk-worms, and have been fully described, and figured in colours, by the late eminent naturalist, Dr. Roxburgh. Both exceed the coehtott silk-worm in size, the first especially; and the silk of which, pears to be more valuable than that of the other kind. The Tusselt sills is found in such abundance in Bengal, and the adjoining provinces, as to have afforded the natives, front time immemorial, all ample supply of a most durable coarse silk, which is woven into a kind of cloth, called Tusseh doot'hies, and much worn. .

The caterpillar, when full grown, is about four incites in length, and bulky in proportion; its colour is green, with a lateral stripe of yellow, edged with red; when ready to spin, they envelope themselves in two or three leaves of the Jujube tree, the vegetable upon which they teed. These leaves form all exterior envelope, which serves as a basin to spin the cocoon in, which is then supended, by a thick silk cord, from the branch of the tree. It remains nine months it; the pupa or chrysalis state, and three months in that of the egg and caterpillar. The insect, when produced, expands to the breadth of fire or six inches, and those of the female to eight inches. They immediately escape.

The worms feed on the trees, and are watched day and night, to guard them against birds. The natives of India pretend that these worms cannot be domesticated. The durability of the silk woven from it, is astonishing. The arrindy silkworm is the *Bombyx Cynthia* of the naturalists. It is peculiar to the interior of Bengal, and is reared in a domestic state, as the connnoli silk-worms. The food of the caterpillar consists entirely of the leaves of the common castoroil plant. It is about three incites long when full fell ; the colour pale green. The cocoon is white or yellowish, of a very soft and delicate texture, about two inches long, anti three in circumference.

The insect remains in the pupa state but twenty days. The filaments of the cocoon are so delicate, that it is impracticable to wind off the silk; it is, therefore, spell like cotton, and woven into a coarse kind off white cloth, apparently of a loose texture, but of incredible durability ; the life of one person being seldom sufficient to \ Near out a garment made of it. The coverings of palanquins are made of this silk. Mr. Glass, British Army Surgeon, sent some of it to his friends in England, and, upon being shown to some silk manufactures, they gave it as their opinion, that it could be made into shawls, equal to any received from India. Another species is called Jarroo, the cocoons of which, are slum ill the coldest month, viz: January; the sills is of a darker colour.

The males, when hatched, invariably fly away, but the females remain oil the asseen trees, upon which the worms are placed to feed. These are not impregnated by the males bred along with them, which fly away : hilt, in tell or twelve hours, another flight of males arrives, and impregnate the females, which deposite their eggs on the branches. The natives are able, to retain part of the Jarroo cocoons for seed, which they hang out on the asseen trees, when the proper season for the moth arrives.

Du Halde mentions that in the province of Chautong, there is found a species of silk in great quantities on trees and in the fields, which spun, and made into a stuff called Kient-chow. This silk is the production of little insects much like caterpillars, which do not spin cocoons, but very long threads ; and, being driven about by the wind, hang upon the trees and bushes, and are gathered for use. The staff is much coarser than that made of silk spun in houses. The worms are wild, and eat indifferently the leaves of the mulberry and other trees ; they are of two kinds, one is much larger and blacker than the common silk-worms, and are called Tsouen. Kien; the other, the Tyan-Kien, are much smaller. The silk of the first is of a reddish gray, that of the other is darker. The stuff made of these materials is between both colours; it is very close, does not fret, is very durable, washes like linen, and, when good, receives no damage by spots, even though oil be spilt on it. Dr. Robertson informs us, that the nature and productions of the wild silk-worms are illustrated at greater length in the large collection of Memoirs on China, and by Pere de Maiila, in his voluminous history of China.

It was to the same species of silk, to which Pliny refers, when he mentions "the stuff made from a white downy substance, combed by the Seres from the leaves of trees, which differed from the wool-bearing trees (cotton) of the island of Tylos in the Persian gulf."

Don Luis Nee observed on certain trees growing in Chilpaneingo, Tixtala, &c., in South America, ovate nests of caterpillars, eight inches long, which the inhabitants manufacture into stockings and handkerchiefs. Great numbers of similar nests, of a dense tissue, resembling Chinese paper, of a brilliant whiteness, and formed of distinct and separable layers, the interior being the thinnest, and very transparent, were observed by Humboldt, in the province of Mechoacan, and the mountains of Santarosa, at a height of 10, 500 feet above the level of the sea, upon various trees. The silk of these nests, which are the work of social caterpillars, was an object of commerce, even in the time of Montezuma; and the ancient Mexicans pasted together the interior layers, which may be written upon without preparation, to form a white glossy pasteboard. Handkerchiefs are still manufactured of it in the Intendancy of Oaxaca.

In the first volume of the Transactions of the American Philosophical Society of Philadelphia, is a paper by the late Moses Bartram, of Philadelphia, in which are recorded some experiments in propagating caterpillars from cocoons, found on the black haw, alder, and wild crab tree. He did not attempt to reel silk from either those he found wild, or from others which were formed in his own house ; but subsequent and recent trials to produce a continuous thread from various native cocoons, have repeatedly proved abortive. The Rev. Mr. Pullein, of England, has indeed recorded, that from the pod of the moth called "isinglass" by Madam Merian, and which he received from Pennsylvania, he produced a thread of twenty single fibres, which bore a weight of fifteen and a-half ounces; while the thread of the common silk-worm, of the same size, always broke with fifteen ounce.

Mr. Abbot also says, that he had heard of the cocoons of the *Bombyx Cecropia* having been carded, spun, and made into stockings, which washed like linen ; but that the insect will riot bear confinement. It feeds on the leaves of the cherry tree. On the outside of the cocoon, the web is coarse; the inside is covered with silk like a silk-worm cocoon.

Miss Rhodes could not succeed in winding any silk from one of the native cocoons, which site received from South Carolina. Madam Hwmbert had some coarse strong silk from cocoons of native wild worms in Louisiana, but, although the cocoons were larger than those of the foreign worms, yet the quantity of silk was less than that produced by the latter.

6

Luna and Polyphemus Moths

Luna moth (*Tropaea luna*) has been rendered unnecessary by the coloured plate, so that it will suffice simply to mention that the beautiful shade of green is of very rare occurrence among our larger moths, and that no other has the long, graceful "tails" on the hind wings, a characteristic which adds greatly to the beauty of this insect. This moth does not seem to be very abundant anywhere, but when once seen will long be remembered on account of its great beauty. The green and yellow colours are evidently very closely related, because either one may, to a greater or less degree, replace the other, so that some of the moths have quite a strong, yellowish tinge.

One of our common swallow-tail butterflies (*Iphiclides ajax*) possesses a very similar green colour in its wings, but does not seem to show this tendency to replace the green by yellow. On the wings are four eyes pots which are also found in Polyphemus. These are remarkable in that they are transparent in the center. This clear area in Luna is quite small, while in Polyphemus it is about as large as the entire eye spot of Luna. The legs are brown and coloured like the front edge of the fore wings. The hairs on the body and at the base of the wing are very long and are white or yellow. The wing expanse ranges from three and three-fourths to five and one half inches.

During April or May the mother moth lays her dark-brown or chocolate-coloured eggs upon hickory, walnut, beech, oak, and a few others of our forest trees. The limited number of food plants is doubtless one reason for the rarity of the moths, as compared with such a common and almost omnivorous larva as Cecropia. A single moth may lay about one hundred eggs, which are smaller than those of Polyphemus. These hatch in about ten or fifteen days, the larva making its escape by eating a circular hole in the shell. Occasionally a young larva may be seen crawling about for a short time, carrying upon its head or tail the empty shell.

The adult larva is about three inches long, of a delicate pate green, a colour very difficult to preserve in the dead larva. Those on the plate have lost this delicate green and have become yellow, but show the form perfectly. This larva is very much like that of Polyphemus, but may be distinguished from it by possessing a longitudinal pale yellow lateral line, which is not found in Polyphemus. Since the cocoon is quite thin and contains but little silk, it is, considered of but little value. This cocoon is spun among two or three Waves, and is about two inches long. Some authors claim that the cocoon falls to the ground with the autumnal falling of the leaves; others that it transforms on the ground among the fallen leaves. The cocoon is quite similar to that of Polyphemus, but not so firmly attached when fixed to a stem. The moths emerge in April and May, there being only a single brood in the north, while there are two in the south.

The colour of the cocoon seems to be influenced in some way by the kind of food eaten by the larva.

Cocoons made by larva which have been fed on hickory leaves have a darker colour. In the true silk worm moth this same influence has been noticed; larvae fed upon the vine producing red cocoons, on lettuce emerald green cocoons, while those fed upon white nettle produce yellow, green or violet cocoons. It is necessary in order to procure these results, that the larvae be fed upon the mulberry till about twenty days before the formation of the cocoon.

Polyphemus – The life history of this native silk worm (*Telea polyphemus*) is by far the best known, because many years ago it was very carefully studied with the hope that it would prove an important silk insect. This hope unfortunately has not been realised. The moths, as shown by the plate, are really beautiful; the large eye spots on the hind wings contributing much towards this effect. The transparent, window-like centers in the eye spot are also of quite rare occurrence among our moths. These transparent areas do not possess the very minute scales found on the other parts of the wing.

Almost all of the wonderful variety of colours found in the wings of butterflies and moths are due either to colouring matter in these scales, or to the breaking up of the white light by minute lines on these scales, such as are seen in the play, of colours on a soap-bubble. These fine lines on the scales are only on the upper side, and are about one-sixteen-thousandth of an inch apart.

The eggs of Polyphemus are very much flattened, about the size of those of Cecropia, and are deposited on leaves and twigs singly or in small groups. These hatch in about ten days and usually in the morning. The young larva often devours the shell which a few moments before afforded it shelter. This larva feeds upon oak, hickory, apple, maple, elm and a variety of other trees, and thus has a larger range of food plants than the Luna larva. The rate of growth is prodigious, as has been shown by Mr. Trovelot.

When the larva hatches it weighs about one-twentieth of a grain; in ten days it weighs one-half of a grain, or ten times its original weight; in twenty days it weighs three grains, or sixty times its original weight; when a month old it weighs thirty-one grains, or six hundred and twenty times its original weight, and has consumed about ninety grains of food; after fifty days it weighs two hundred and seven grains, or over four thousand times the original weight. At fifty-six days the larva has eaten eighty-six thousand times its original weight in food! It is therefore not surprising that these larvae can often be easily detected upon trees by the large number of leaves which they have devoured.

To provide for this great change in size, the larva moults five times, but the time between these moults is not always the same; there is usually about ten days between the first four moults and about twenty between the fourth and fifth. The larva stops eating a day before the moult, spins a few threads upon the leaf to which it attaches its hind legs, and waits for the transformation, which usually takes place in the afternoon. The larva, when mature and ready to spin its cocoon, is about three inches long. It is sometimes influenced in its colour by the food plant; the normal larva being of a golden green, although it has been known to show more yellow colouring when found on red maple.

A short time before beginning its cocoon the larva ceases to eat and selects a place for its cocoon. These cocoons are usually found upon the ground among the leaves, but are frequently attached to twigs. After about a half days work the larva spreads over the inside of the cocoon a gummy, resinous substance, which binds together the threads. After four or five days more of almost continuous work, another coating is smeared over the inside, which renders the cocoon practically air-tight. The silk fibres become considerably finer as the cocoon nears completion and the supply of silk begins to run low. For this reason the inner layers of the cocoon are only about half as strong as the outer ones.

The larva, as the supply of silk diminishes in the silk glands, becomes perceptibly reduced in size. It has been estimated that the larva, in attaching the continuous thread of its cocoon, makes two hundred and fifty-four thousand back and forward movements. The cocoons are very strong and dense, of a dirty white colour and generally coated with a white powder, the female being the larger.

There is but a single brood in the north, while in the south there are two. In order to see if the pupa needed air, Mr. Trovelot sealed up some cocoons over winter in shellac, but the moths emerged in due time after being in an airtight space for nine months. He also delayed the emergence of the moth till twenty-one months after entering the cocoon by placing it upon ice.

The silk in the spinning glands before it is spun is a clear, transparent fluid. These glands seem to be of excessive size when compared with that of the larva, since, when fully expanded, they reach the great length of twenty-five inches, or about eight times the length of the full grown larva. These glands are paired, one being found on each side of the body, are considerably folded and taper at each end. The ducts leading from the anterior end of the glands unite to form a single duct which opens below the mouth. The thread is double, being really composed of two different fibres, one from each gland, as may be shown by separating them. The silk in these glands is prepared and sold as silk "gut" to anglers.

On account of its, transparency when in water, it becomes invisible and thus aids in deluding the wary fish, who does not see any connection between the line and the baited hook. The "gut" is prepared as follows: Larvae which are ready to spin their cocoons are cut open and placed in strong vinegar for eighteen hours; the glands are then taken out, stretched and dried in the shade.

Six or eight days after beginning the cocoon, the larval skin is moulted and the real chrysalic or pupal stage begins. This stage normally lasts till the following spring or summer. A few days before the time of emergence a pair of glands which open into the mouth become very active and secrete an acidulated fluid which escapes and wets the fore end of the cocoon, causing the resinous material binding together the fibres to become soft. Even cocoons sealed up in shellac and starch have been dissolved by this fluid, and thus the moths have been able to escape.

When the cocoon has become sufficiently soft, the moth pushes its way between the fibres, but in doing so often breaks some of the threads, thus making the silk of such cocoons useless, for commercial purposes. The moth at the time of emergence, with its folded and crumpled wings, is quite a forlong-looking object. These wilted wings soon begin to fill up with fluids from the body, which is very large at this time.

In some cases, the fluid is driven into the wings with so much force that they swell up, and if such a wing is punctured, thus allowing some of the fluid to escape, the mature wing will be of a smaller size than one from which no fluid has been lost. It must be remembered that it is possible to inflate a butterfly or moths wing, because the wings of insects are not composed of a single layer, but are sacs of two layers which are closely applied. It is thus possible to split the wing into upper and lower halves, but this can only be done at the time of emergence, when these two layers are not so firmly cemented together as they are in a few hours after emergence.

The enemies of Polyphemus are numerous. Birds prey upon the larvae, in addition to numerous parasitic insects which are very similar to those which destroy Cecropia. The cocoon itself is not a complete protection because rats, and squirrels plunder them. We thus see that the life of even an insect is full of dangers, and that it is really a wonder that so many are able to become mature and reproduce. The silk-worm moths are excellent illustrations of what is called complete metamorphosis in insects. An insect like the grasshopper, when it hatches from the egg, is very much like the adult insect in its general form and appearance; the most evident difference being the lack of wings.

An insect which shows such slight changes in its growth to maturity is said to have an incomplete metamorphosis. It is incomplete in the sense that the change is not of a very radical nature. But in the case of the silk worm moths, and moths and butterflies in general, the larva which hatches from the eggs has not even the most superficial resemblance to the adult insect, the fully developed moth. This necessitates a complete change or metamorphosis in the form and structure of the insect before it can become mature. This great change is accomplished during the quiet pupal stage in the cocoon. Because the pupa is apparently passive when viewed from the exterior, one must not conclude that it is so internally; far from it; the digestive organs of the larva must be completely made over from those of a chewing leaf eater to those of a moth which can only take liquid food.

Giant Silkmoth Collection

The Saturniids or giant silkmoths are among the largest and most spectacular of American moths and favorites of collectors (Fig. 1). All are beautifully and brightly coloured and the largest of the group may have a wingspan that exceeds 6 inches. What insect collector has not wanted to catch or rear the lovely luna moth or the uncommon Columbia moth!

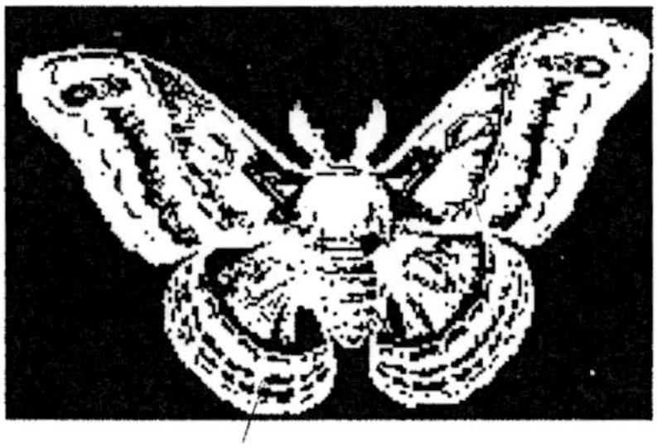

Figure 1

Silkmoths can be collected at any stage of their development but it is most rewarding to collect them in winter when they are in their cocoons and awaiting their emergence. This is good winter sport and needs only a minimum amount of effort, and you certainly will get the best specimens. Several species of silkmoths spin their cocoons on trees or shrubs which become conspicuous after leaf drop. Hunting cocoons is simply just knowing where and what to look for. Several silkmoths such as the Cecropia and Polyphemus have a wide range of food plants; a few like the Columbia moth feed only on one tree species (Table 1).

Some skill in recognizing their food plants in their winter apparel is useful. Once located, the cocoons, because of their size and shape, stand out against the narrower twigs and branches. A good cocoon hunter, in order to find the most desirable trees or shrubs, must think like the female moth. When she is ready to lay her eggs she searches for shrubs or young trees and prefers isolated ones or those at the outer edges of woods, marshes, and bogs where maneuvering is easiest. Your most profitable searches then should be along fencerows, in backyards, along railroads and roadsides, and in openings where there are small clumps of trees and shrubs.

Figure 2

Cecropia moth cocoons are often over 3 inches long and the largest produced by native silk moths. They are made of tough silken threads and always attached lengthwise to the twigs or branches of the host.plants (Fig. 2). Those high on the plant tend to be longer and more compact while those spun near the ground are more bulky and somewhat loosely spun. Cecropias, probably because they have such a wide variety of food plants, frequent urban areas where lilac, maple, fruit trees, and birch are planted.

Figure 3

Promethea moths are usually the most common of the silkmoths and their cocoons resemble curled leaves that didn't fall from the tree in autumn (Fig. 3). Actually, each cocoon is wrapped in a leaf that the larva pulled about itself as it spins its silken home. The silk covers the leaf petiole and firmly anchors the leaf and cocoon to the twig. It is about 1-2 inches long and hangs vertically from the twig. Prometheas prefer wild habitats to urban ones so hunt for them along roadsides, railroads, and forest edges for best results. When you find one there should be others nearby. "Where there's one there's four; and where there's four, there's more," is a good saying quotes Dave Eppelheimer, an avid cocoon collector. Its not uncommon to find 6 or more on a single small tree or shrub.

Polyphemus moth cocoons are about 2 inches long but more rounded than those of Promethea moths (Fig. 4). The Polyphemus larva camouflages its cocoon by incorporating leaves into its structure and then fills the fibers with a fluid that becomes white and chalky. Like the Cecropia it readily feeds on plants in urban situations, so look for these cocoon on shrubs and small trees in yards, vacant lots, and parks. Also search beneath these food plants as some cocoons are spun on the ground and others fall off the plant.

Figure 4

The thin papery cocoon of the highly coveted luna moth is similar in form and size to that of the Polyphemus but contains much less silk. Although it is spun among the leaves it falls to the ground in autumn and thus is more difficult to find than the latter. If you search diligently beneath its food plants you may be rewarded with a cocoon or two.

Table 1. Food plants of common silkmoth larvae (in order of preference)

Silk Moth	Food Plant
Cecropia moth	cherry, maple, boxelder, elderberry, willow, lilac
Polyphemus moth	maple, boxelder, basswood, plum, oak, walnut
Promethea moth	cherry, lilac, spicebush, ash, basswood, plum
Luna moth	walnut, hickory, butternut, oak, beech, birch, willow
Io moth	cherry, willow, elm, boxelder, aspen, maple, corn
Columbia moth	larch (tamarack)
Cynthia moth	ailanthus (tree of heaven)

Columbia moth cocoons are oval, about 1-1/4 to 2 inches long and comprised of tough brown silken threads. silver streaks often mark its sides. Like the Cecropia, its nearest relative, the Columbia attaches its cocoon to the twig (Fig. 5) or occasionally to the trunk of its host plant. However, it is a rarer insect and feeds only on larch (tamarack) so you must search for it in localities where larch thrives such as in or near acid bogs, and on young trees up to 25 ft. tall. Most cocoons are 2-6 ft. from the ground and toward the tips of the branches. Occasionally it spins its cocoon on shrubs beneath the larches. Larch cones sometimes make it difficult to spot the cocoons. The best time to collect is in autumn shortly after larches lose their needles but before the birds and other predators locate and destroy the cocoons.

Figure 5

The cocoon of the Cynthia moth is spun up in a single leaf and resembles that of the Promethea moth. This species is far less common, however, and feeds only on the ailanthus or tree of heaven which is planted in urban areas. Introduced in eastern North America in 1861 from Asia for silk production, it has now become widespread in the Northeastern part of the United States. The Io moth cocoon occurs on the ground as does that of the Luna moth, and though possible to locate, you must do considerable hunting. You can collect the larva, which feeds on numerous hosts, or the adult much more readily.

Silkmoths in autumn- or winter-collected cocoons will not emerge unless given a "winter period". You can overwinter cocoons in an unheated garage or in the refrigerator from October to late winter. If then brought indoors they will emerge. If you plan to have the moths mate and lay eggs and then rear the larvae, be sure to overwinter them until at least late April or until their food plant has sufficiently leafed-out.

Silkmoths can also be collected in the larval and adult stages. The conspicuous larvae or caterpillars of some attain 3 inches long and many are armed or adorned with tubercles, spines, or hairs. All can be handled except the Io moth which has stinging hairs on its body. When touched it feels like stinging nettles, so collect it without handling it. Blacklights attract most adult silkmoths so they can be collected this way during their summer evening flights.

7

Molecular Genetics of the Silk Moth, *Bombyx Mori*

The molecular genetics of silkworm genome is poised on the edge of a new era. The large collections of carefully maintained silkworm mutations and practical breeding stocks are valuable resources that have only begun to be exploited to their full potential. An intensive genomics programme for the silkworm should enhance its usefulness for strain improvement and aid in the development of tools for targeted knockouts.

Most silk moth species belong to either the family of Bombycidae or Saturniidae, in the superfamily of Bombycoidea. Phylogenetic studies based on interspecific hybridisation, chromosome pairing behaviour and gene structure comparison suggest that the present day domesticated silk moth, B. mori, is closely related to B. mandarina, the wild silk moth. Cytological investigations of inter-specific hybrids of Japanese B. mandarina (n = 27) and B. mori (n = 28) have revealed the formation of a trivalent with one chromosome of the former and two chromosomes of the latter suggesting that one mandarina chromosome might have split into two during the history of domestication.

As in most Lepidoptera, chromosomes of the silkworm are holocentric, i.e. they possess centromeres throughout the chromosome body. This was shown by the persistence of chromosome fragments after irradiation during cell division, the dispersion of microtubule attachments, the absence of recombination nodules in females, the presence of supernumerary chromosomes in some close relatives of the silkworm, and the chromosome pairing behaviour and fertility of interspecific hybrids of two Saturniid species with diverse chromosome number.

B. mori chromosomes are highly condensed and appear dot-shaped at most meiotic and mitotic metaphase stages. Diffused centromeres and lack of special features make them difficult to identify individually. This has resulted in limited application of modern cytogenetic tools for genetic and molecular studies. The silk moth, B. mori has been used as a model system for formal genetic studies since the discovery of Mendelian inheritance at the turn of the century because of its large size, ease of rearing in the laboratory, and economic importance in silk production.

The well-developed genetics of this species includes more than 400 mutations which have been mapped to 28 linkage groups or chromosomes. In addition, the existence of hundreds of geographic races and genetically improved strains used for silk production which differ not only in Mendelian traits but also in the complex of quantitative traits such as body size, feeding duration, thermal tolerance and disease resistance, makes them amenable to systematic analysis using modern genetic tools.

Research on silkworm was initiated in Japan as early as 1900 under the patronage of the sericulture industry which rapidly promoted it as a valuable model system for fundamental studies. In 1906, Toyama reported the utilisation of the principles of hybrid vigour to enhance viability and yield by crossing different silkworm genotypes. The phenomenon of sex-linkage in B. mori was first reported by Tanaka, shortly after its initial report in D. melanogaster. Many genetic phenomena such as maternal inheritance, parthenogenesis, polyploidogenesis, radiation-induced translocations and complex loci were reported in the early 1930s.

Together with the many inherent merits of the silkworm, the growing collection of world genetic stocks, encompassing 3000 genotypes, places B. mori second only to Drosophila as an insect model for genetic studies. The formulation of artificial diets as early as in 1960, freed the silkworm rearing from the use of fresh mulberry leaves and helped to expand the study of Bombyx to academic laboratories. The accumulation of information on the nutritional requirements of the silkworm led to the formulation of artificial diet.

The composition of artificial diet has been gradually improved and there are only small differences in the rate of growth and development of larvae reared on mulberry leaves or artificial diets. Many of the research findings have already provided important insights into the understanding of basic biological processes. In silkworm, females are heterogametic (ZW) and males are homogametic (ZZ) in sex chromosome constitution. The change of ratio of sex chromosomes to autosomes has no effect on sex determination and gynandromorphs, that is, sexual mosaics, rather than morphologically blended intersexes were produced in such studies, suggesting that sex determinants are active on both Z and Wchromosomes.

The triploid and tetraploid silkworms which carry ZZW and ZZZW chromosome constitution invariably develop into females, suggesting that female determinants are possibly localised on W-chromosome. In light of detailed investigations on sex-determination mechanism in Drosophila, it may be worth investigating the primary signals in Bombyx sex determination and the degree to which the underlying genetic and molecular mechanisms are conserved. In silkworm, there appears to be no dosage compensation. This has been demonstrated using an anonymous Z-specific transcript and Bm kettin, homologous of the X-linked Drosophila kettin gene.

In both cases, the level of transcripts in males was two times greater than that in females unlike dosage compensated species such as mammals, in which one copy of the homogametic chromosome pair (XX) is inactivated in somatic tissues by heterochromatinisation. On the other hand, in many lepidopteran species, as well as in other taxa with heterogametic females, there is ample evidence that it is the W-chromosome that becomes inactivated. In silkworm and other lepidopteran insects, heterochromatic body is shown to be present only in females in somatic tissues such as silk glands, malphighian tissues, epidermal cells, etc.

The sex chromatin body has been correlated with the W-chromosome in endomitotic tissue such as silk gland, artificially induced polyploidsand in mutants carrying translocations between the W and various autosomes of silkworm. In light of the findings that W-chromosome is strongly female determining and there is no dosage compensation in Bombyx, it would be of interest to investigate whether heterochromatin of W-chromosome specific sequences shows any features with well-studied phenomenon of X-chromosome inactivation in mammals.

The identification of W-chromosome specific sequences would facilitate further investigation of this process at the molecular level. B. mori was the first insect shown to have an interspersed pattern of repetitive and non-repetitive sequences typical of mammalian genomes. Molecular characterisation of the repetitive elements from B. mori has revealed the presence of transposable elements typical of the Drosophila genome as well as retroposons typical of the mammalian genomes.

The B. mori genome harbours abundant retroposons (SINEs, for short interspersed nucleotide elements repeats) like Bm and Bm elements which represent 5 to 10% of the total genomic mass, similar to the Alu and Alu-like elements found in the mammalian genomes. Interestingly, the fruitfly genome is devoid of such SINE elements. Unlike the mammalian genomes, which contain only a non-LTR (Long Terminal Repeat) LINE-1 elements (Long Interspersed Nucleotide Elements), the Bombyx genome contains several diverse elements some of which contain long terminal repeats like Pao and Mag and others like RBm, RBm and BMC which lack long terminal repeats similar to the many LTR (Copia, Gypsy)- and non-LTR elements (RDm, RDm, Jockey, F, G and I) described in Drosophila.

Bombyx genome also harbours DNA-mediated mobile elements such as mariner similar to fruitfly genome. Telomeric DNA motifs have been characterised from the silkworm and are made of a pentanucleotide repeat (TTAGG) characteristic of many insects instead of the usual hexanucleotides (TTNGGG) found in vertebrates. A similar telomeric repeat appears to be present in wild silk moths and other insects, although not in Drosophila, whose telomeric region contains a complex array of longer and more diverse sequence elements which are also found in heterochromatin.

A number of sequenced B. mori genes encoding proteins with diverse functions are now available (Table 1). In this regard, the silk gland and the chorion provided effective model systems to help unravel the molecular basis of gene expression. The silk gland has long been used as a model for studying gene regulation and cell differentiation. An attractive feature of this tissue is its simple construction with a succession of territories comprising a fixed number of cells each expressing a specific set of silk protein encoding genes.

Table 1. Selected genes cloned from Bombyx mori

Cloned genes	Encoded products
Enzymes	
Sorbitol dehydrogenase	Sorbitol dehydrogenase
Trehalase (cDNA)	Trehalase
Glycyl-tRNA synthetase	Glycyl-tRNA synthetase
cdc2 and cdc2-related (cDNA)	Cell cycle kinases
Homeoproteins	
Bmftz-F1	Fujitaratzu-related
Bmantp	Antennapedia
BmUbx	Ultrabithorax
Bmcd1	Caudal
Bmabd-1	Abdominal
Bmen	Engrailed
Bmin	Invected
BmScr	Sex comb reduced
Bmdef	Deformed
Hormones and receptors PTTH	Prothoracicotropic hormone
DH/P-BAN	Diapause hormone/ pheromone
	Biosynthesis activating neuropeptide
Bombyxin A, B, C and D	Insulin related peptides
EcR	Ecdysone receptor

BmEcR-A and B1 (cDNA)	Ecdysone receptor isoforms
Immune proteins	
BmLeb	Lebocin
GNBP (cDNA)	Gram negative bacteria-binding protein
FPI-F (cDNA)	Serine protease inhibitor
BmAtt	Attacin
MORICIN (cDNA)	Moricin
CecB1 and 2	Cecropin
BmLyz	Lysozyme
Pro-PO (cDNA)	Prophenol oxidase
Regulatory proteins	
SCF (cDNA)	Super coiling factor
SGF-1	Silk gland cell regulatory protein
POU-M1/SGF3	Sericin gene regulatory protein
BmGATA a and b	GATA-like regulatory proteins
RNAs	
rDNA	Ribosomal RNA
tRNAgly	Silk gland glycine tRNAs
tRNAala	Silk gland alanine tRNAs
Structural proteins	
BmColl (cDNA)	Collagen-like peptide
BmBtub (cDNA)	B-tubulin
A3	Cytoplasmic actin
A4	Cytoplasmic actin
A1 and A2	Muscle actins
ErA and ErB	Early chorion proteins
A and B	Middle chorion proteins
HcA and HcB	High cysteine late chorion proteins
PCP-2	Pupal cuticle protein
H-Fib	Heavy fibroin chain (silk protein)
L-Fib	Light fibroin chain (silk protein)

P25	Fibroin-associated peptide
Ser-1	Sericins (Silk gum protein)
Ser-2	Sericins (Silk gum protein)
ESP	Egg storage protein
SPI, SP2	Larval haemolymph storage proteins

The single silk gland cell accumulates up to 400,000 to 800,000 haploid genomes by endomitosis, the highest known polyploidy ever reported. The silk proteins, fibroin and sericins are two distinct families of proteins. Fibroin is a complex of three associated polypeptides encoded by single genes, (the heavy and light chains) which are linked by disulphide bonds and p25, a co-secreted chaperone loosely attached to the two others. Sericins are the products of two distinct genes whose precursor mRNAs are the target of differential splicing leading to 4 and 2 sericin isoforms from the genes Ser 1 and Ser 2, respectively.

Functional assays for regulatory elements of chorion genes in follicular cells have been elegantly demonstrated using the heterologous transgenic system of Drosophila, and follicular cell regulatory factors and their genes are now well characterised. Homeobox genes, Bombyx ultrabithorax (BmUbx), Bombyx abdominal-A (Bmabd-A) and Bombyx Abdominal-B (BmAbd-B), which constitute a part of the E-pseudoallelic complex that specifies the identity of body segments, have been cloned and characterised.

From the initial molecular analysis of the homeodomains the E-complex has been shown to be analogous to the Bithorax complex (BX-C) of Drosophila. However, some mutations of the Ecomplex reveal one interesting aspect that is different from those in Drosophila. E-mutant phenotypes such as E/E and E/+

cause shifts in one direction on the dorsal side and in the opposite direction on the ventral side. Such independent determinations of the dorsal and ventral sides is not observed in Drosophila.

Differences between the Bombyx E-complex and the BX-C of Drosophila show that the structures and functions of the cisregulatory and coding regions outside the homeodomains of the Bombyx Ecomplex may be different from those of Drosophila which remain to be investigated. Another homeotic complex, Nc-locus which harbours homeotic genes for thoracic segments, Bombyx Antennapedia (BmAntp) and sex comb reduced (Scr), which specifies identities of maxillae, labial and prothoracic segments has also been characterised. These two genes of Nc-locus lie in close proximity of approximately 100 kb and are located on the sixth chromosome at 1 cM distance from the E-complex.

The Nc-locus corresponds to the Drosophila ANT-C. The close proximity of Nc and Ecomplex in the sixth chromosome is similar to ANT-C and BX-C in Drosophila; likewise the adjacent location of Bm Antp and Scr genes within approximately 100 kb is as in Drosophila, where they are positioned at about only 70 kb distance. Earlier attempts at producing transgenic silkworms have resulted either in transient expression or in rapid degradation and extensive rearrangement of the injected sequences.

Recently, the lepidopteran mobile element, Piggy bac, has been successfully put to the germline of silkworm (Pierre Couble, University of Lyon, France, pers. commun.). In another novel approach the recombinant Autographa californica nuclear polyhedrosis virus (Ac NPV) carrying chimeric fibroin light (L) chain green fluorescent protein (GFP) gene has been successfully used to target the chimeric gene to the Lchain region of the silkworm genome by homologous recombination. The integrated gene has been shown to stably transmit through generations.

Genetic Stocks of Silkworm

The silkworm genetic stocks maintained around the world comprise geographical races, inbred lines, and mutants that carry numerous morphological, developmental, behavioural and biochemical features. It is estimated that more than 3000 silkworm genotypes are being maintained in Asia and Europe. The genetic diversity found in the present day inbred lines and elite stocks is mostly a result of cross breeding of geographical races which display a variety of qualitative, quantitative and biochemical traits.

Broadly, four sets of geographical races have been identified: Japanese, Chinese, European and Tropical, which in addition to many visible characters, differ widely in qualitative and quantitative characters that affect silk yield. The genotypes of temperate origin produce high quantities of rated silk while the tropical races are poor producers, but hardy and able to survive under adverse climatic conditions. The genetic stocks of the silkworm comprise more than 500 mutants for a wide variety of characters (serosa colours; larval and adult integument colours, skin markings and body shapes; cocoon colours and shapes; physiological traits such as diapause, number of larval moults and timing of larval maturity; food habits and biochemical features such as digestive amylase, blood amylase, blood and egg esterases, larval integument esterase, alkaline and acid phosphatases, haemolymph proteins; silk production and fibroin secretion; homeoproteins and body plan determination, etc.) (Figure 1).

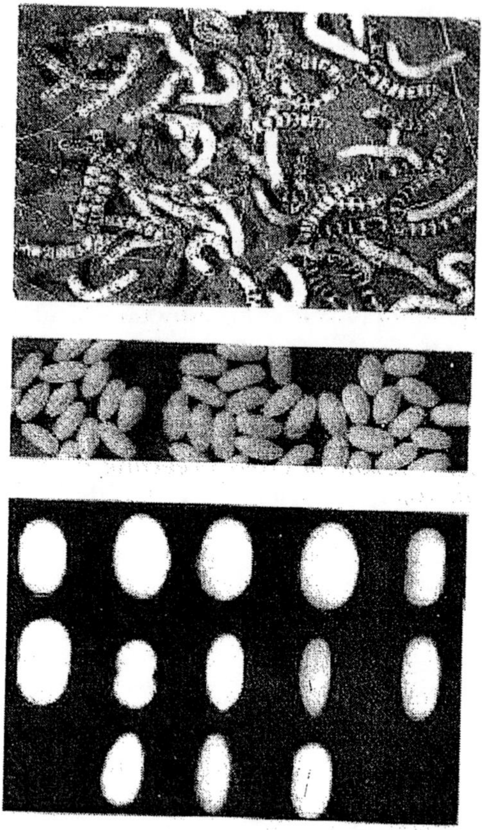

Figure 1. Genetic resources of silkworm. An array of larval mutants (upper); different ecotypes and inbred lines showing various cocoon shapes, sizes and colours (middle and lower).

B. mori, the only lepidopteran with well developed genetics, proves itself an ideal organism to apply the emerging molecular technologies to construct high density map. Such a map has manifold applications: positional or map-based cloning, localisation of complex traits, comparative studies on genome evolution, marker-assisted breeding and selection and development of tools for fingerprinting. Further, information generated from B.

mori can significantly benefit many economically and ecologically important lepidopteran species which include many wild silk moths and the most destructive agriculture pests.

The haploid genome of silkworm comprises 560 million base pairs per silk glandcell, which is approximately 3.5′ the size of the D. melanogaster genome (140 million base pairs) and one-sixth the size of the human genome. Given the 560 Mb of the silkworm genome, more than 3000 molecular markers would be required to place one marker within every 250 kb, an average insert size which could be packaged in YAC (Yeast Artificial Chromosome) or BAC (Bacterial Artificial Chromosome) if one plans to use the genetic map for positional cloning using YAC or BAC Library.

In order to pursue this scale of molecular map one requires high volume marker technology. The recent explosion of knowledge on the use of molecular techniques and remarkable progress made in the large-scale genomics of mammalian systems have resulted in the high volume marker assays that promise precision, speed, and cost effectiveness. These methods have accelerated the progress in high density mapping of various genomes, map-based cloning of genes of interest for which no prior information was available, identification of quantitative trait loci (QTL) for economic characters and utilisation of molecular markers for 'marker assisted selection' and estimation of genetic diversity. Such an integrated approach would form the basis for the analysis and manipulation of the silkworm genome, and provide important raw material for comparative genomics and germline transformation.

Selection of suitable parents and synthesis of appropriate genetic populations are important steps in the molecular mapping strategy. The two parental strains, Cand p50 have been used for molecular linkage map. The former is a diapausing strain (bivoltine - 2 generations per year) and the latter a non-diapausing strain (polyvoltine - 5-6 generations per year) and they exhibit high phenotype diversity for such complex characters such as size, growth rate, diapause, morphology, nutritional requirements, general vigour and cocoon properties suggesting that considerable polymorphism exists at DNA level.

These strains have undergone a high degree of inbreeding and are relatively homozygous. Both the strains are of Chinese origin. The Fprogeny raised from a single pair mating of Fsibs have been used for molecular map construction and various molecular markers (restriction fragment length polymorphism (RFLP), random amplified polymorphic DNA (RAPD), simple sequence repeat polymorphism (SSR) and inter-simple sequence repeat (ISSR)) are being integrated to this common Fmapping population.

It is common to use backcross population for linkage mapping in B. mori because of lack of crossing over in females. These backcrosses have differing advantages depending on the sex of the informative Fparent. If the Fis female, rapid detection of linkage is possible because of the complete absence of recombinants but the estimation of map distance is not possible. If the Fis male, the estimation of map distance is possible because of crossing over, but the detection of linkage is more difficult.

Clearly, one approach to molecular mapping is to conduct the eight backcrosses obtainable by all possible combinations of the initial cross (C108 ´ p50 or p50 ´ C108), sex of the parent, and recurrent strain for the backcross (C108 or p50). This would permit mapping of co-dominant loci or dominant loci originating from either strain making data analysis straightforward. For maximum efficiency in the initial phases of linkage mapping, an Fcross if used, would yield eight-fold savings in labour with only about a two-fold decrease in information content.

Molecular Markers

Molecular markers have several advantages over the traditional phenotypic markers that are available in silkworm genetic stocks. They are unaffected by environment, detectable in all stages of development and ubiquitous in number covering the entire genome. Many techniques have been employed in silkworm to reveal the genome-wide polymorphism. These are RFLP, RAPD, SSR and ISSR anchored PCR. A preliminary RFLP map has been developed in silkworm using 52 progeny of an Fcross from a pair mating of two parental genotypes, Cand p50.

The RFLP probes included 15 characterised single copy sequences, 36 anonymous sequences derived from a follicular cDNA library and 10 loci corresponding to a low copy number retrotransposon. The RFLP map covers a total recombination length of 413 cM dispersed over 15 linkage groups. Out of the 15 characterised single copy sequences, 10 (which have already been localised on the conventional linkage map using biochemical and morphological variants) could be correlated with the molecular map. Although RFLP analysis offers several advantages, it requires large quantities of DNA, is labour intensive, time consuming and expensive.

The advent of PCR technology increased the approaches available for the detection of DNA polymorphisms, which offer effective alternatives to the hybridisation methods of RFLP. The PCR-based methods meet the wide spectrum of requirements for large-scale genetic studies since they require only small quantities of DNA, avoid DNA blotting and radioactivity and are amenable to automation.

RAPD marker technology involves the amplification of genomic DNA template primed by a 10 base-oligonucleotide primer of a random sequence but with a minimum guanidine-cytosine content of 50%. A low annealing temperature (37-40° C) is maintained throughout the cycles (35-40 cycles). Polymorphisms occur as indicated by the presence or absence of a specific amplification product among individuals. A medium density linkage map of RAPDs has been constructed using 102 F individuals of the same parents (p50 and C) used for the RFLP map.

The screening of 320 decanucleotide-primers resulted in 243 clear polymorphic products between Cand p50, of which 168 bands followed Mendelian inheritance. The MAPMAKER program sorted 168 bands into 29 linkage groups and 10 unlinked loci covering the 892.4 cM map distance. Recently, a high density linkage map containing 1018 genetic markers representing all 27 autosomes and the Z-chromosome has been constructed. Most of the markers were RAPDs with primer pairs in combinations of 140 commercially available decanucleotides.

The use of random primer pairs has greater advantages in map construction because of exponential increase of combinations of primer pairs using a limited number of primers and higher reproducibility when compared to using single primer. Considering the 560 Mb genome size of the silkworm, the average interval of markers works out to be ~560 kb and the relationship between physical and genetic distances is ~250 kb/cM. This value is intermediate between the honey bee (52 kb/cM)and the fruit fly (575 kb/cM). These values agree well with the idea that crossing over rates are inversely correlated with chromosome size, because average chromosome sizes are 11.1 Mb for the honey bee, 17.1 Mb for the silkworm and 47 Mb for the fruit fly.

The overall distance of Yasukochi's map is nearly 2 cM/marker. The routine application of RAPD analysis for genetic mapping has limitations. First, the efficiency of amplification of fragments using decamers at low annealing temperatures (37°C) is very sensitive to reaction conditions and may be inconsistently reproducible in different experiments. This is particularly problematic if one is interested in integrating mapping data from different crosses by comparison to common reference markers. Second, RAPD markers are inherited as dominant markers. Lack of allelism is a disadvantage as the heterozygous and homozygous loci are not distinguishable.

Three criteria define ideal polymorphic loci:

(i) they should be highly polymorphic;

(ii) they should be easy to identify and type; and

(iii) they should be amenable to automation.

Microsatellite based genetic markers, which are distributed across most of the eukaryotic genomes, fulfil these criteria. They have proved to be extrordinarily efficient means to pursue genetic mapping. Microsatellites are short stretches of DNA which consist of an array of simple di-, tri-, or tetranucleotide repeats such as (CA), (GA), (GTA), (ATT), (GATA). A unique oligonucleotide on each side of the repeat region is chosen for the production of a primer pair for each of the microsatellite array (locus).

Allelic variations are based mostly on differences in the number of tandem repeats in a microsatellite array. Thus, the only way in which alleles can be distinguished is by measuring the total length of the microsatellite array. This is most readily accomplished through PCR amplification of microsatellite itself along with a small amount of defined flanking sequences on both sides with designed primer pairs for each locus. A higher level of cost effficiency is achieved by combining two or more loci for simultaneous analysis through multiplex PCR. By using the fluorescent labelled primers and automated sequencers, the entire process of microsatellite analysis can be automated.

Recent studies have shown that the silkworm genome is abundantly interspersed with CA/GT and GA/CT repeats. The (GT)repeats occur at every 49 kb while (CT)repeats occur, on an average, at every 104 kb in the silkworm genome. These frequencies compare favourably with the well-characterised mouse and human genomes where intensive maps of microsatellite markers have been accomplished. Initial studies show that the primers for the flanking microsatellite loci revealed distinct allelic differences between different strains of silkworm and inherit in Mendelian pattern.

In the thirteen silkworm populations analysed using 14 microsatellite loci, the number of alleles ranged from 3 to 17 with PIC values of 0.66 to 0.90. The microsatellite markers can thus provide high volume markers for intensive mapping of the silkworm genome. As many microsatellite loci could be multiplexed based on allelic sizes in an automated sequencer, the microsatellite markers could be mapped quickly using the appropriate silkworm mapping population. Efforts are now underway to utilise these markers to integrate with the existing RFLP and RAPD maps.

Examination of conservation of microsatellites among distantly related saturniid silkmoths such as Antheraea mylitta, A. pernyi, Philosamia cynthia ricini, A. yamamai and A. polyphemus and patterns of cross-species polymorphism revealed that almost one-third of the 60 Bombyx microsatellite loci tested are conserved in the heterologous species. These results show that microsatellite loci provide another potential source of markers for comparative genome work in lepidoptera.

As PCR technology is finding increased use in genetic analysis, more variations of this technique are emerging. One such variation takes advantage of the ubiquitous distribution of di, tri and tetra nucleotide repeats in the silkworm genome using primers designed for the repeat regions to amplify the inter repeat genomic regions. The primers consisted of (CA) or (GA) repeats extended into the 5¢ or 3¢ termini by 2 to 4 purine or pyrimidine residues to initiate PCR amplification of genomic segments flanked by inversely oriented, closely repeated sequences.

The radioactively labelled primers are used for PCR amplification of template genomic DNA. The amplified DNA when resolved under nondenaturing polyacrylamide gel electrophoresis reveal information on multiple genomic loci in a single lane. Such markers are dominant and are shown to inherit in Mendelian fashion. Although most ISSR loci are dominant, rather than co-dominant, they offer several advantages over RFLPs and RAPDs for genotyping, the major one being the rapid production of a large number of markers in a cost-effective manner.

The ISSR markers offer higher reproducibility due to the use of longer primers and higher annealing temperature than those used for RAPDs. Due to these advantages we believe that the ISSR technique has great potential in silkworm breeding and germplasm evaluation. The ability to screen so many polymorphisms in a single assay makes the ISSR-PCR a technique of choice for a large-scale screening of silkworm germplasm and makes it useful to augment the marker resources for the silkworm genome mapping programme.

The banded krait minor satellite DNA Bkm-2(8) containing a 545 bp sequence consisting mainly of 22 GATA repeats was also used as a probe to reveal polymorphisms in different silkworm races. Although polymorphism level was relatively lower than SSR anchored PCR, the Bkm probe revealed very clear fingerprint profiles. Such DNA fingerprint-based information will be useful to estimate genetic diversity of silkworm germplasm. Two bacterial artificial chromosome (BAC) libraries have been constructed from the two B. mori reference strains, p50 and C.

The libraries contain a total of 36,864 clones and are approximately 5.8 and 3.2 genome equivalents, respectively. The average insert sizes in the libraries are 134.5 kb and 120.8 kb, respectively. The previously reported close linkage between the Bombyx homologues of the invected (Bm in) and engrailed (Bm en) genes was confirmed in the newly constructed BAC contigs that contained both. This is the first report on large-insert libraries of non-dipteran insects. Such libraries are available only for D. melano-gaster.

With the ongoing efforts on construction and integration of high density linkage maps, the availability of BAC libraries and mutant genetic rêsources of the silkworm, map-based cloning will be feasible within a few years, and light will be thrown on molecular mechanisms that control quantitative trait loci (QTL), various mutant phecotropes. These libraries will be used for map-based cloning of genes and projected large-scale genomic sequencing projects. The construction of a complete physical map has been initiated by a hybridisation method using independent ESTs as probes.

In a complementary approach based on PCR, RAPD markers derived from the high density maps and STSs are being used to construct contig islands encompassing known silkworm genes and gene complexes such as Hox genes; this will provide information about the chromosomal organisation of regions likely to be conserved in other species of insects.

8

Silk from Caterpillars and Spiders

Silk thread is synthesized from digested mulberry leaves, it is technically an animal fiber rather than a plant fiber. Another fascinating animal fiber is derived from the webs of spiders. In fact, the world's strongest natural filament comes from the silk strands of a spider's web, specifically orb weaver spiders of the genus *Nephila* (family Araneidae).

The golden orb's web can run from the top of a tree 20 feet (6 m) high and up to six feet (2 m) wide. Unlike the fragile webs of other spiders, the golden orb's web can last for weeks and even months. The silk of *Nephila* species is so strong that it can trap small birds, which the spider doesn't eat. Trapped birds often destroy the web by thrashing around. To avoid such damage, the spider often leaves a line of insect carcasses on its web (like the safety strip on glass doors), or builds smaller barrier webs around the main web.

Matted and twisted webs of *Nephila maculata* are used by South Sea Islanders for various kinds of bags, fishing lures and traps. Female spiders are induced to build nests on bamboo frames which are then used as fishing nets. In the Solomon Islands, the spider web is collected by winding it around sticks to make large sticky balls which are suspended just above the water. Needle fish are lured to jump out and get entangled in the ball. In Southeast Asia, people make a net by scooping up the web between a stick bent into a loop.

Spider webs have also been used as a bandage to stop blood flow from an injury. The silk of golden orb spiders is almost as strong as Kevlar, the strongest synthetic fiber which is drawn from concentrated sulphuric acid. If spider silk could be manufactured, it would have thousands of uses, including parachutes, bullet-proof vests, lightweight clothing, seatbelts, and light but strong ropes. It could also be used for sutures in operations, artificial tendons and ligaments. It has been estimated that a solid strand of silk from this spider the diameter of a pencil could stop an airliner in flight! This unsubstantiated claim was reported on a TV nature program.

Studies are now being conducted to have genetically engineered plants produce fluid polymers that can be processed into silk. Spiders are not used commercially to produce silk fabric because silkworm moth caterpillars produce twice as much silk and are much easier to culture. Up to 150 meters of silk can be collected from a single individual of *Nephila clavipes.* It would take approximately 415 spiders to make a square yard of cloth. The same number of silkworm larvae could make twice as much silk, and they don't eat each other. All you need is a plentiful supply of mulberry leaves.

Close-up view of a female golden silk spider (Nephila clavipes) in her web. The orb web of this species may be over one meter in diameter. This species is also called the golden orb spider because of the golden tinge of the silken web. This large spider is easily recognised by the conspicuous tufts of hair on the femora and tibiae of leg pairs #I, #II and #IV.

Spider Silk

Many types of spiders construct specific types of silken web to trap prey. These come in a variety of forms, from the familiar orb web and cobweb to sheets, funnels, tubes, ladders, and even a single thread. A single spider may produce more than a half-dozen different kinds of silk for different purposes, some sticky, some not. One kind of silk, for example, is used by many web-spinning spiders to wrap up prey after it initially becomes trapped in the web. This immobilises the prey and makes it safer for the spider to approach and bite the potential prey.

Ogre spiders (family Deinopidae) hold a small web in their front legs, and use it as a net to snare insects. Bolas spiders (Mastophora species, in the family Araneidae) have abandoned the "traditional" web in favor of a silk line with a ball of sticky glue at the end. These spiders exude a chemical which mimic certain female moths by secreting pheromones similar to the ones the moths use to attract mates. The male moths are drawn to the spider, which throws the glue ball at the moth and "reels" its victim in like an angler hauling in a fish.

Spiders use silk in mating activities, too. Male spiders construct a "sperm web" in which they secrete sperm from their gonads, so that it can be sucked up into the pedipalps. These modified mouthparts are not directly connected with the abdominal gonads. Males of many species seem to locate females by following draglines laid down by the females. These draglines seem to contain chemical signals telling the males that an appropriate female laid down the silk. Some males also use silk to tie up the females as part of a courtship ritual.

Mating can be dangerous for male spiders because females are often larger and may consider the male a better meal than a mate. These silk "nuptial veils" may give the male a little extra time to escape after mating. All female spiders wrap their eggs in silk egg sacs, often using several different kinds of silk, some leathery for protection from the outside environment, some soft and fluffy to cushion the eggs. Most spiders constantly lay down a "dragline" as they walk, which acts as a safety line if they need to jump or drop away from a potential predator, for example. Some species hang their egg sacs in or near their webs, while others carry their egg sacs around with them.

Many larger spiders find shelter in burrows they dig in the soil. These are usually lined with silk and may be capped with lids called "trap doors" which are made of silk and other materials such as soil. Many smaller spiders construct various retreats and "sleeping bags" out of silk, as well. Small spiders can move through the air by "ballooning," in which a long strand of silk attached to the spider lifts the spider up in rising air currents. By this method, spiders are able to disperse to new habitats, even to remote islands, and habitats can be recolonised rapidly after a disturbance, such as a fire that might kill most spiders living in a prairie.

Spiders can only balloon if they are small (less than about 5-8 mm long), so many spiders balloon only as small immatures. Some small spiders such as sheet-web spiders (Linyphiids), though, can balloon as adults, too. Silk is clearly one big reason for the great success of spiders in so many habitats.

The cobweb spinning spider, of course, had already become legendary by inspiring the oft-defeated Robert the Bruce, the Scottish king by showing him how to succeed. His exclamation "If at first you do not succeed, try, try again!" remains with us still, as also the spider's ability to spin webs. Only, the latter precedes Robert

Bruce by at least 100 million years. It has been estimated by scientists that spider silk is at least five times as strong as steel, twice as elastic as nylon, waterproof and stretchable.

Genetic studies affirm that the basic silk recipe dates back to the earliest orb weavers—a class of cobweb spinning spiders that evolved 100 million years ago, and suggest that it leaves very little to be improved upon. Without hot temperatures, high pressure or harsh chemicals, spiders are able to produce a steady stream of silk, one of the toughest fibres known.

Nature

In the case of silk spider, large scale forming is ruled out as they do not like living near each other. Also, it appears silk proteins are complex molecules and are difficult to decode. According to researchers, the secret lies in the spinning rather than what is spun. Within the silk gland, the silk proteins are dissolved in a highly concentrated solution that is half water and half proteins. From the silk gland the protein is squeezed out of a duct. The proteins fold up in rod shapes and are pulled out of the duct by knobby appendages known as spinnerets and they unwind, lock together and solidify into silk fibre.

Recently, Nexia Bio Technologies in Quebec successfully genetically engineered goat's milk to produce spider silk. The company spliced silk, making genomes of its goats configured so that the genes would act on only the mammary glands of lactating female goats. The silk proteins are purified out of the milk into a whitish viscous liquid and then spun into fibres. These fibres can then be used for applications such as medical sutures and fishing line—stronger than nylon, but environment-friendly as it decomposes over time.

Spiders and insects have been able to spin silk for at least 380 million years. It has been estimated that the strength of spider silk is at least five times as strong as steel, twice as elastic as nylon, waterproof, and stretchable. Scientists have long envied the strength and elasticity of spider's silk but have been unable to synthesise it. The dragline silk of the Golden Orb-weaving Spider is the most studied in scientific research. Dragline silk is the fiber from which spiders make the scaffolding of their webs. Of late, demonstrations of cloning and expression of silk fibroin protein, including spider dragline silk, represent an exciting scientific opportunity and technological challenge.

Spider silk is a natural polypeptide, polymeric protein and is in the scleroprotein group, which also includes collagen (in ligaments) and keratin (in nails and hair). Spider silk genes encode proteins composed of iterated peptide motifs, and the consensus sequences are multiply repeated throughout the silk proteins. Patterns of 'Ala-rich' blocks and 'Gly-rich' amorphous blocks provide both strength and elasticity, and such biomaterials might therefore be valuable for industrial and medical purposes. Although spider silk was initially produced in animal cells, mass production from animal cells is expensive and the quantity produced is limited, mainly due to the requirement for large fermenters, sterile conditions, and expensive media.

In order to overcome these factors, plants (tobacco, potato, *Arabidopsis*, and soy somatic cells) have been evaluated for the production of spider silk proteins. These studies demonstrated the feasibility of plant based silk-like protein (SLP) production. However, high yield and low production costs remain a challenge. Jianjun Yang and co-authors, from Central Research Development and

Crop Genetics Research, Du Pont de Nemours & Co, USA, report high yield recombinant SLP production through protein targeting in transgenic *Arabidopsis*.

The researchers used a synthetic *DP1B* gene for producing SLP, achieving variable accumulation levels through protein targeting into a special cellular components. DP1B is a synthetic analogue of spider dragline silk protein, which can be spun to form silk fiber. An average DP1B accumulation level in plant cells below 1.5% of total soluble protein has been previously reported. However, Yang et al. showed that ER and vacuole targeting in seeds and apoplast, and ER targeting in leaves, can dramatically increase SLP accumulation.

Many transgenic plants engineered with the ER targeting mechanism were able to accumulate DP1B SLP in their seeds to a level greater than 15% of total soluble protein. A high yield of SLP was achieved by utilising *DP1B-8P*, which is a plant-optimised, synthetic 8-mer *DP1B* gene that encodes a 64-kD DP1B SLP. Targeting of SLP to the designated cellular components was enabled by adding a 21-amino acid sporamin signal peptide, 16-amino acid sporamin propeptide, and 4-amino acid ER retention peptide to appropriate positions on the DP1B molecule.

To express the resultant DP1Ba, DP1Be, and DP1Bv fusion proteins in *Arabidopsis*, expression plasmids pGYV501, pGYV502, and pGYV503 were introduced into *Arabidopsis thaliana* by an *Agrobacterium*-mediated floral transformation method. Healthy T1 transgenic plants were grown in soil and T2 seeds were harvested from individual T1 plants to represent independent transformation events. Later, T2 seeds were germinated on selective plates to obtain the next generation. All T2 transgenic seedlings originating from the same parents were grown together in soil. T3 seeds were collected from these plants and pooled together to represent independent transformation events.

Standard molecular assays were performed to confirm transgene expression in the independent transformation events. DP1B production yield was calculated based on DP1B content and relative to total soluble protein content in the extract. The transformation was confirmed by PCR amplification of sporamin signal peptide coding sequences from the genomic DNA. Leaf protein extract made from a pGY401 *Arabidopsis* plant was used to show control levels of unmodified, cytosol-accumulated 64-kD DP1B.

The results of immunoblot assay showed that in leaf tissues of *Arabidopsis*, apoplast and ER targeting are suitable approaches for DP1B production, but the vacuole targeting is not. Results also showed that accumulation of the unmodified DP1B in leaf tissues of pGY401 plants ranged from 0.01% to 1.7% of total soluble protein (TSP), with an average yield of 0.3%, while DP1Ba accumulation in leaf tissue of pGYV501 plants was from 2.1% to 8.5% TSP, with an average yield of 4%. DP1Be accumulation in leaf tissues of pGYV502 plants was from 0.3% to 6.7% TSP with an average yield of 1.5%.

Vacuole targeting did not result in accumulation of correctly-sized DP1Bv molecules. This comparison of accumulation levels between the non targeted and targeted DP1Bs revealed that DP1B productivity in leaf tissues was increased 13-fold by targeting to apoplast and 5-fold by targeting to ER lumen. Apoplast targeting is the best approach. Characterisation of seed-specific DP1B fusion proteins showed accumulation of unmodified DP1B in pGY411 seeds ranging from 1% to 1.4% TSP with an average yield of 1.2%.

The accumulation of the ER-targeting DP1Be in pGYV512 seeds ranged from 1.8% to 18% TSP (avg. 9.4%), and the vacuole-targeting DP1Bv in pGYV513 seeds ranged from 4.8% to 8.2% TSP (avg. 6.5%). Thus a comparison between the targeted and non-targeted DP1Bs suggested that the ER and vacuole targeting significantly enhanced DP1B productivity in seeds by 7.8- and 5.4-fold, and at the highest levels, up to 15-fold increase was observed in ER targeting.

Importantly, DP1B transgenic plants were heritable after one and two cycles of sexual reproduction, as confirmed by transgene insertion and expression studies in T2 leaves or T3 seeds, and the accumulation level of DP1B fusion protein was stable. Thus this research demonstrates the highest level of DP1B accumulation among all protein targeting strategies. It was concluded that combining seed-specific synthesis and ER targeting is the best method of production. Though *Arabidopsis* is not a good candidate for commercial production, the approach may be tried in an industrially important crop that supports the large-scale plant based production of DP1B SLP.

9

Silk Industry

Sericulture is a labour-intensive industry in all its phases, namely, cultivation of silkworm food plants, silkworm rearing, silk reeling, and other post-cocoon processes such as twisting, dyeing, weaving, printing and finishing. It provides employment to approximately 58 lakhs persons, most of them being small and marginal farmers, or tiny & household industry mainly in the hand reeling and hand weaving sections.

Cultivation of sericulture is not very widespread being practiced regularly in contiguous districts in the three southern States of Karnataka, Andhra Pradesh and Tamil Nadu; in the NER; in the tribal areas of Jharkhand, Chhattisgarh, Andhra Pradesh, Orissa; and in Jammu & Kashmir and West Bengal. The cocoons and silk processing industry has traditionally existed in clusters in areas not necessarily coinciding with the sericulture areas in Karnataka, Andhra Pradesh, Tamil Nadu, West Bengal and Bihar. Reeling of silk is more widespread, with different clusters/traditionally producing distinctive designs and weavers, mainly in saris.

Technology upgradation in the silk industry has been slow, and has been adopted mainly in Karnataka. India has the distinction of cultivating all the four commercially known varieties of silk, namely, Mulberry, Tasar, Eri and Muga.

World Raw Silk Production

The World Raw Silk production (mulberry and non-mulberry) about 125629 MT (2004- P) mainly from two countries, China and India. China leads the world with silk production of 102560 MT or 81.6% of the produce. India ranks second in respect of world raw silk production. It is this position, as one of only two major silk producers in the world, and from its employment potential, that sericulture and silk derive their importance in the Indian textile map. Policy decisions are defined mainly by these two considerations.

Cultivation Area and Domestic Raw Silk Production

India's raw silk production is 16500 MT (2004-05), 88.61% (14620 M.T.) of it is mulberry and 11.39% (1880 M.T.) Vanya silks. The area reported by States as cumulatively planted under mulberry is 1.72 lakh hectares (2004-05); and the production of cocoons is 120027 tonnes.

Productivity of Raw Silk

As seen from the table below, though, area under sericulture has fallen- mainly because of drought in AP and Karnataka in the last three years, as also due to price fluctuations, the productivity per ha. has been steadily improving. The main reasons behind increase in production is- improved silkworm races and hybrids, better technologies in rearing and reeling, evolution of higher leaf yielding mulberry varieties.

Year	Mulberry Hectarage	Production of raw silk (Tons)				Per Ha. Productivity (In Kg.)	
		Mulberry		Non-Mulberry	Total		
		Cocoon	Raw Silk			Cocoon	Raw silk
1998-99	270069	126566	14260	1284	15544	468.64	52.80
1997-98	282244	127495	14048	1188	15236	451.72	49.77
1999-00	227151	124531	13944	1270	15214	548.23	61.39
2000-01	215921	124663	14432	1425	15857	577.35	66.84
2001-02	232076	139616	15842	1509	17351	601.60	68.26
2002-03	194463	128181	14617	1702	16319	659.15	75.16
2003-04	185120	117471	13970	1772	15742	634.57	75.46
2004-05	171959	120027	14620	1880	16500	698.00	85.07

Bivoltine Silks

With Japanese technology and cooperation, Central Silk Board (CSB) has recently been able to evolve & popularise Bivoltine silkworm races, which can yield raw silk of international standards. With these races, provided there is concommitant reforms in the marketing and processing of cocoons, India can hope to expand its sale of domestic raw silk beyond its own borders. The X Plan target for bivoltine raw silk production is 6700 tonnes.

Silk Imports

(i) *Demand Supply Gap*: The domestic production of raw silk is not adequate to meet the domestic and export demand. It is estimated that against the demand of around 26,000 ton per annum the domestic production is around 16500 tonnes. The gap of nearly 9500 tonnes in demand is mainly on account of the fact that high-grade quality mulberry raw silk is not being produced in the country to the extent required by the industry. This quality of mulberry raw silk is basically required in the power-loom industry, for export purposes, and to some extent in the handloom industry for warp purposes. To meet the demand of exporters, the Govt. has allowed the import of raw silk under O.G.L. as per WTO compatibility.

(ii) *Raw Silk Imports*:-The quantity and value of raw silk imported during the last five years is indicated as under:

	Raw Silk	
Year	Quantity (in ton)	Value (Rs. in Crores)
1997-98	2346	218.33
1998-99	2824	259.36
1999-2000	5018	412.74
2000-2001	4713	475.15
2001-2002	6808	624.73
2002-03	9054	647.15
2003-04	9258	628.41
2004-05	7948	607.21

Import Duty

Duty has been brought down progressively from 44.04% to 30% Justification for enhancement of customs duty on silk yarn and fabric in the Union budget 2004-05.

Years	Customs Duty
2000-01	44.04%
2001-02	40.40%
2002-03	35.20%
2003-04	35.20% on raw silk and 20% on silk yarn and silk fabric
2004-05	30% on raw silk and 30.192% on silk yarn and fabric.

During 2004-05 customs duty has been increased on silk yarn and silk fabric from 20% to 30.192%. This has been done for the following reasons:-

(i) The duty structure was inverted with lower duty on value added items and higher duty on raw material. With this budget the duty on all silk items in the chain are now at par with each other at 30% duty. After decrease in customs duty on fabric the trend of imports has shifted to fabric and the quantum of imports in respect of fabric has increased very heavily. Import of silk fabric during 2002-03 stands at 3285 MT which was 839 MT in 2000-01 with respective value of Rs.27.05 crores in 2000-01 and Rs.212.78 crores during 2002-03. During 2003-04 and 2004-05 the silk fabric import were 3208 MT (Rs.406.85 crores) and 3700 MT respectively (Rs.571.61 crores).

(ii) in order to protect the interests of domestic farmers and reelers as heavy imports bring instability in the domestic market and farmers and reelers are not able to get remunerative price for their produce which are about 26 lakh families.

Anti-Dumping Duty on Import of Raw Silk

The policy towards sericulture has been protectionist till recently. Since 1/4/2001 when the import of raw silk has been put under O.G.L. as per WTO compatibility, the import of silk yarn and fabric showed an increasing trend in comparison to the corresponding period of last year, which resulted in fall in the prices of raw silk in the domestic silk industry during the period 2001 to June, 2003 and disrupted the domestic market.

Antidumping investigations were undertaken by Directorate General of Anti-Dumping & Allied Duties

(DGAD&AD) on a petition by the reelers affected by the falling domestic prices. The designated authority recommended provisional imposition of antidumping duty on landed goods so as to raise import prices to US$ 33.19 per kg, pending final determination, on all imports of Mulberry Raw silk (not thrown), 2A grade and below, originating in, or exported from Peoples' republic of China.

This was notified for 6 months by Govt. on 2nd January, 2003. The Directorate General of Antidumping announced the final determination on 3rd July, 2003 at US$ 27.97 per kg. on all imports of mulberry raw silk (not thrown) of international grade 2A and below, originating in or exported from the People's Republic of China. This has helped to stabilise domestic prices.

Petition for Anti-dumpling Import of Silk Fabric from China

Weavers Association's have filed a petition before the Directorate General of Anti Dumping and Allied Duties alleging dumping of silk fabrics by China and requesting for imposition of Anti Dumping Duty on exports of silk fabrics from China. At this juncture, CSB in association with the various silk Power Loom Weavers Associations situated in Karnataka, Gujarat & Uttar Pradesh have filed a Case on 29th September 2004, against the dumping of cheap Chinese silk fabric with the Director General of Antidumping & Allied Duties (DGAD), Ministry of Commerce (GOI).

The designated authority of DGAD has issued a Gazette Notification on 18th May 2004 initiating investigation process. A team from Director general of Anti-dumping and Allied duties have visited Bangalore on 7th-8th October 2005 to evaluate the extent of injury occurred to the silk industry. The process of investigation is under progress for imposition of Interim duty on Chinese silk fabrics.

Exports

(i) Export Trend

Exports of Indian silk products comprise mainly natural silk, fabrics, made-ups, ready-made garments, silk carpets and silk waste. Indian Silk exports have grown during last few years, rising from Rs.1250.55 in the year 1998-99 to Rs.2421.98 crores in the year 2000-01. However, in the year 2001-02 and 2002-03 export of silk goods showed a declining trend ie Rs.2359.56 crores and Rs.2294.05 crores respectively. The export of silk goods during 2003-04 was Rs.2779.19 crores (US$ 604.7 million). The silk goods exports during 2004-05 was 2879.56 crores (US$ 640.90 million).

(ii) Action Taken for Export Enhancement

1. The Government is providing assistance to the Indian Silk Export Promotion Council, Mumbai for undertaking various export promotion activities such as participation in international fairs, organising generic promotion stalls in domestic fairs like TexStyles India, Publicity in foreign trade magazines, publication of magazine "Silk India" and colour forecast card for domestic silk manufacturers, dissemination of overseas trade information among exporters, etc.
2. Government has rationalised value addition/Input-output norms specified under the EXIM Policy extended to exporters, the facility of duty free import of raw material under the Advance Licensing Scheme, import of capital goods at concessional rate of duty for export products.

3. For Technological up-gradation of the industry loan at 5% point less than applicable rate, under the Technological Up-gradation Fund Scheme for the Textiles sector is available inter alia to the silk sector.
4. Import duty on silk machinery has been reduced to 10%.
5. A production programme of high quality of bivoltine silk to improve quality of Indian silk to international standards and its productivity & cost competitiveness is being pursued. Besides great thrust is being given for promotion of non-mulberry silks in the existing areas of North East and Jharkhand, expanding it to other areas like Uttaranchal, Himanchal Pradesh, Bihar etc.

Programmes are being implemented to improve technologies at all stages of the process of silk development, and diversify products with inputs for improvement of looms and designs such that they may expand the export share of the country.

Export Data

The exports of silk goods during the last five years is indicated as under:

Year	(Figure in Rs. crore)*	Million US $
1998-99	1250.55	297.04
1999-2000	1755.55	404.97
2000-2001	2421.98	530.21
2001-2002	2359.56	495.29
2002-2003	2294.05	474.08
2003-2004	2779.19	604.70
2004-05	2879.56	640.90

* Including silk carpets and silk RMG.

Net foreign exchange from silk, it may be noted, is reduced to two thirds to half this figure, due to rising imports.

National Textile-2000 (NTXP 2000)

The new Textile Policy of GOI (NTxP) has incorporated the following directive for silk: Focus will be on achieving international standard in all varieties of silk. Steps will include:

- — Improving Research & Development and the effective transfer of technology at all stages;
- — Considerably improving the production of non-mulberry varieties of silk
- — Augmenting efforts for the spread of bivoltine sericulture;
- — Encouraging clustering of activities of reeling and weaving and strengthen linkages between the producers and industry;
- — Periodically reviewing the import policy for raw silk taking into account the balanced interests of the sericulturists as well as export manufacturers.

X X-Plan

Targets for X-Plan

i) Production:

In M.Tons

Silks	Incremental	Cumulative
Mulberry-Bivoltine	6400	6700
Mulberry Multivoltine	650	17450
Tasar/Oak Tasar	115	450
Eri	550	1700
Muga	55	150
Total	7770	26450

ii) *Exports*: Exports are envisaged to grow at 15% reaching a level of about Rs.3200 crores by end of X-Plan.

iii) *Employment*:- Creation of livelihoods is projected at 60.03 lakhs.

iv) *Implementation of Plan Schemes*: During X Plan an outlay of Rs.450 crores has been allocated for development of Sericulture. During 2002-03 & 2003-04 an amount of Rs.89.29 crores and Rs. 88.64 crores has been spent respectively. During 2004-05 an amount of Rs.85.85 crores was released. For 2005-06 the allocation is Rs.110 Crores (including 30.00 crores for North East).

Current Issues

1) *Legislation and market reforms*: The traditional sericulture States have enacted legislation that require all reelers and rearers to register themselves, physically bring all their produce to the designated market yard and to none other and to compulsorily participate in the open auction conducted by the State. The presence of a large number of small rearers with low capacities has led to the production of small lots of cocoons of varying quality, thus driving away corporate sector from setting up large cocoon processing units and also the existing system has inhibited the weaving sector from promoting better technology oriented reeling units as the system of open auction in the silk exchanges does not allow them guaranteed access to the output from these reeling units, nor any advantage in terms of price.

In order to bring the necessary changes in the existing legislation, policies, management system and schemes legislative amendments in the respective State's Acthave been proposed by the CSB & circulated to the states of Karnataka, AP & TN. The issues have been taken up with the State Governments mainly the Southern Sericulture Sates and have been pursued at the level of Secretary (T) and Minister of Textiles but response of the State Governments is not forthcoming.

(2) *Launching on Line Auction and Futures Trading in Cocoon and Silk yarn*: Central Silk Board has launched futures trading in cocoon and raw silk at the National Commodity and Derivatives Exchange (NCDEX) in Mumbai on 20th January, 2005. The futures contracts on mulberry cocoon and raw silk have been jointly designed by the Central Silk Board and NCDEX. The sericulture industry would benefit by futures trading through better price discovery, quality linked pricing and the dissemination of market information.

(3) *Amendment to the Central Silk Board Act, 1948 to provide for regulation of the production and sale of silkworm seed*: Sericulture provides livelihood to an estimated 25 lakh small and marginal farmers in the country and occupies about 2.3 lakh hectares of irrigated and rainfed land. Altogether the silk industry generates employment for 58 lakh persons, and an output of about 16500 MT of raw silk. It contributes an estimated Rs.6000 crores to the GDP.

Sericulture has traditionally been carried out under a regulated trade regime with restrictions on imports, exports, movement of goods between states, and highly controlled markets. Since 2001, globalisation has exposed the industry to several risk factors and has necessitated changed responses.

Among these, the most urgent is the need to regulate production, supply, distribution, trade and commerce in quality silkworm seeds. With this objective, Ministry of Textiles have introduced "Central Silk Board (Amendment) Bill, 2005", in Rajya Sabha. "The Central Silk Board (Amendment) Bill, 2005" as introduced in Rajya Sabha has been referred by the Hon'ble Speaker to the Parliamentory Standing Committee on Labour which visited Karnataka on 09.11.05 and heard the farmers, reelers and weavers. Further action like notifying the Bill etc., is awaited.

(4) *Mid term appraisal of the X plan schemes*: Sericulture, which falls within the Village and Small Enterprises Sector, is important for the creation of sustainable gainful employment and rural incomes, providing a non-agricultural commercial crop with a high level of rural income and for supplying the raw material to industry largely based in rural areas that has a significant and growing export.

The targets for the sector were fixed in terms of production of mulberry and non-mulberry raw silk, creation of employment and enhancement of exports. The planned budget of Rs.450 crores was provided to implement the following schemes to assist the achievement of these targets:

1. R&D/Transfer of technology/training
2. Seed Support & Technical Assistance/HRD
3. Catalytic Development Programme
4. IT initiatives
5. Quality Certifications Systems (central scheme)
6. UNDP-Assisted Fibre Development Programme

Reasons for Shortfall

The major development schemes has been done in consultation with the State Governments and other implementing agencies. The objective of this appraisal was not only to assess pace of progress and identify short comings but also to assess whether it was proving effective in tackling the major issues identified at the start of the Xth plan:-The short falls in production during 2002-03 and 2003-04 were due mainly to:

1. The drought conditions which prevailed during last 2 years in the Southern States which are major silk producing States.
2. Increasing imports of raw silk and silk fabric at low prices from China since 2001 adversely impacted the domestic markets thereby distracting farmers and other stakeholders from sericulture activities as it has reduced the remunerative gains for sericulturists, releers and weavers.
3. Obsolete reeling and weaving technology limit productivity and profit margins. With reeling and weaving confined mainly to the small and tiny sectors, the beneficiaries lack capacity to invest in upgraded technology. The availability of credit to the rural industry sector was identified as a bottleneck. Despite continuous discussions with the traditional sericulture states especially Karnataka to amend their acts regulating the market, the lack of response to the globalising conditions and the continued restraints on sale of cocoons worked adversely for the sector. The production of bivoltine silk was affected mainly by this.
4. The functioning of cocoon markets does not permit reelers to obtain a steady supply of uniform quality

of bivoltine cocoons due to the lack of systems for grading and direct linkages with cocoon growers. Without this reelers are unable to produce a uniform quality of silk and therefore to achieve a gradable silk or get a remunerative price for their produce.

5. Seed supply, which falls 73% in the private sector under license by the State Governments and 20% directly in the State sector is un-reliable for both quality as well as disease freeness. Controls by the State Governments on licensed seed producers are so poor that they are unauthorisedly crossing hybrids and to sell untested, un authorised and un guaranteed seeds to the farmers. In turn their supply of seed cocoons for producing silkworm seeds lack proper grading and sorting.
6. Release of funds by the State Governments to the department is less of a problem in sericulture than in several other sectors, but is by no means always tiny. It is also been seen that in several states, state share has not been provided to meet the requirements of the sector or the requirements projected in their proposals.
7. Appraising the schemes individually it is seen that the scheme for IT initiatives has been slow in taking off because of the lead time required to make a systems study for the IT requirement of the sector (i) an on line spot auction system to cater to the states where markets for cocoons are not functional and (ii) to set up a virtual agricultural extension system through the E-Chaupal concept. Both systems have now been delineated and the requirements of funds has been worked out.

Additional funding would be required for which a proposal under the NCMP is being moved.

8. A Component by Component appraisal of the Catalytic Development Programme (8 sub schemes) has shown that components meant for creation of rural infrastructure have been extremely popular and ahead of the envisaged targets; the demand for support to reeling machinery and the setting up of Common Facility Centre has been slower than expected due to the slump in the raw silk prices and the difficulties of obtaining credit.

Based on the above appraisal, following remedial action has been initiated:-

1. Continuous efforts are being made for increasing production and bringing quality of produce at par with international standards by popularising cultivation of bivoltine mulberry silk in suitable areas, evolution of more productive hybrids and expansion of production of non mulberry silk i.e. tasar, eri and muga by popularising production of these varieties in non traditional areas.
2. Through Research and Technological development reeling technology is being developed and popularised.
3. For regulation of production and supply of silkworm seed, setting up of National Silkworm Seed Regulatory Authority is under process.
4. Setting up of systems for futures trading in silk and on line auction of raw silk and cocoons is being processed in consultation with the relevant authorities.

World Rawsilk Production

Country	1995	1996	1997	1998	1999	2000	2001	2002	2003	2004	% (P)
						Share)					
China	67113	68500	60300	57500	56959	61648	64567	68600	94600	102560	81.64
India	13909	14126	15236	15544	15214	15857	17351	16319	15742	16500	13.13
Japan	3240	2580	1920	1080	650	557	431	394	287	287	0.23
Brazil	2468	2270	2120	1821	1554	1389	1485	1607	1563	1512	1.20
Korea Republic	946	506	272	210	200	165	157	154	150	150	0.12
Uzbekistan	1320	2500	2000	1500	923	1100	1260	1260	950	950	0.76
Thailand	1313	1144	1039	900	1000	955	1510	1510	1500	1420	1.13
Vietnam	2100	1500	1000	862	780	780	2035	2200	750	750	0.60
Others	2967	2766	2117	1572	1250	1952	1692	3814	1500	1500	1.19
Total	95376	95892	86004	80989	78530	84403	90488	95858	117042	125629	100

Note : Figures of India is for Financial year April to March.
Source : 2004- Sericologia 44 (3)& ISCC-2005
P: Provisional
SS: 2212/2005

5. *Antidumping on excessive import of raw silk from China*: Since liberalisation of imports in 2001, import of raw silk has been steadily increasing with low prices. Raw silk imports have increased from 4713MT in 2000-01 to 9258 MT in 2003-04. Increasing imports from 2001-02 to 2002-03 impacted the domestic price of raw silk. Price of imported silk in India fell drastically from US $ 24.50 per kg in May 2001 to US $ 13.50 per kg. in June 2003. This disrupted the domestic market and prices of cocoons and raw silk fell by about 40% due to large volume of import at low prices.

The farmers and the reelers who were the most affected segment approached the Directorate General of Anti Dumping and Allied Duties (DGAD) requesting for levy of anti-dumping duty against China. As dumping of raw silk by China was established, the GOI imposed antidumping duty against the import of mulberry raw silk (not thrown) of international grade 2A and below, originated in or exported from People's Republic of China.

State-wise Raw Silk Production for 2004-05

Unit: Tonnes

State	2004-05				
		Vanya Silk			
	Mulberry	Tasar	Eri	Muga	Total
Andhra pradesh	5084	16	10	-	5110
Assam	9	-	554	104	667
Arunachal Pradesh	neg	-	4	0.1	4.1
Bihar	8	8	15	-	31
Chattisgarh	2.3	120	1	-	123.3
Gujarat	-	-	-	-	-

Contd....

Contd....

Himachal Pradesh	12	-	-	-	12
Haryana	-	-	-	-	0
Jammu & Kashmir	90	-	-	-	90
Jharkhand	Neg.	91	1	-	92
Karnataka	7301.9	-	-	-	7301.9
Kerala	8	-	-	-	8
Madhya Pradesh	14	14	-	-	28
Maharastra	37	5	-	-	42
Manipur	54	3	431	0.1	488.1
Mizoram	5	neg	3	0.1	8.1
Meghalaya	2.5	-	327	5.4	334.9
Nagaland	0.5	0.2	98	0.1	98.8
Orissa	2	33	1	-	36
Punjab	4	-	-	-	4
Rajasthan	0.3	-	-	-	0.3
Sikkim	-	-	-	-	0
TamilNadu	443	-	0.6	-	443.6
Tripura	4	-	-	-	4
Uttar Pradesh	9.5	2	-	-	11.5
Uttaranchal	9	4	0.6	-	13.6
West Bengal	1520.09	26	2	0.2	1548.3
Total	14620.09	322.2	1448.2	110	16500

Despite Antidumping duty on all imports of mulberry raw silk of grade 2A and below from China, the imports of raw silk are on the increase. The antidumping duty is being evaded by China by exporting to India raw siik of Grade2A and above and twisting silk yarn which is beyond the purview of the antidumping notification.

6) *Dumping of silk fabrics into India from China*: Import of silk fabrics from China is also increasing very

> fast against import of 188MT during 2000-01 this has increased to 2824 MT in 2003-04. During 2004-05 import of silk fabrics from China is estimated to 3558 MT. Silk Weavers' Associations have filed a petition before DGAD for levy of antidumping duty against imports of silk fabrics. Directorate General of Antidumping & Allied Duties (DGAD) has issued Initiation Notification NO.14/20/2004-DGAD dated 18th May, 2005 initiating antidumping investigations into the existence, degree and effect of the dumping of the silk fabrics originating in or exported from the People's Republic of China.

At this juncture, CSB in association with the various silk Power Loom Weavers Associations situated in Karnataka, Gujarat & Uttar Pradesh have filed a Case on 29th September 2004, against the dumping of cheap Chinese silk fabric with the Director General of Antidumping & Allied Duties (DGAD), Ministry of Commerce (GOI). The designated authority of DGAD has issued a Gazette Notification on 18th May 2004 initiating investigation process. A team from Director general of Antidumping and Allied duties have visited Bangalore on 7th-8th October 2005 to evaluate the extent of injury occurred to the silk industry. The process of investigation is under progress for imposition of Interim duty on Chinese silk fabrics.

(7) Organisation of 20th Congress of the International Sericulture Commission, 2005: Central Silk Board organised 20th Congress of the International Sericulture Commission, 2005 from 15th to 18th December, 2005 at Hotel Taj West End, Bangalore. International Sericulture Commission, headquartered at Lyon, France is the only inter-governmental organisation of silk producing

countries. The chief objective of International Sericulture Commission is to encourage and promote the development of sericulture and allied activities of sericulture industry in the member countries through technology transfer, collaborative Central Sector and Centrally sponsored Plan Schemes: Following Central and Centrally Sponsored Schemes are being implemented during X Plan period:

Catalytic Developmental Schemes (CDS)

The Central Silk Board in collaboration with the State Governments implements Catalytic Developmental Schemes in the X plan. Some schemes are 100% funded by Central Silk Board, some are jointly funded by CSB and States or 50:50 cost sharing basis, in some schemes beneficiaries i.e. farmers/reelers/NGOs/Cooperatives/private entrepreneur also contribute their share with the CSB's/State share. In the States of NER, including Sikkim, and J&K, Himachal, Chattisgarh, Jharkhnad, Uttaranchal, central share has been increased to 90%.

The CSB as well as State Governments do monitoring of these schemes. State Governments furnish Utilisation Certificate to the CSB against each scheme. The records of the beneficiaries benefited under these schemes are maintained by the respective State Governments. These schemes cover all the four varieties of the silk ie mulberry, tasar, muga and eri. Total budget allocation during the IX Plan for CDS was 89.27 crores and total expenditure was Rs. 47 crores. In the X Plan the allocation to the scheme is Rs.180.98 crores.

Externally Aided Projects

i) Japan International Cooperative Agency (JICA) Assisted Bivoltine Project

The first phase of JICA assisted Bivoltine Sericulture

Technology Development Project (BSTD) was implemented from 1991 to 1997. With the implementation of this project, a comprehensive practical Bivoltine Sericulture Technology suitable to Indian conditions was developed.

The second phase of JICA assistance implemented the Project for Promotion of Popularising the Practical Bivoltine Sericulture Technology (PPPBST) with active participation of Karnataka, Andhra Pradesh and Tamil Nadu from 1997 till 2002. The PPPBST Project activities entailed verification and demonstration of Bivoltine technologies developed in the first phase for the project with selected farmers, training of Indian sericulture technologists, technicians, field staff and farmers. The performance of these Bivoltine races matched international standards, with average yield over 60-80Kg/ 100Dfls, a renditta of 5.5 to 6 Kgs and a consistently high quality 2A-4A grade raw silk.

Encouraged with the results, the third phase of the Project on "Strengthening of Extension System for Bivoltine Sericulture" was commenced from August 2002 for a period of five years with a target to establish suitable extension and seed supply systems for promotion of bivoltine in the selected TSCs in Karnataka, Tamil Nadu and Andhra Pradesh and in CSB. The project proceeded on schedule during the year.

As of today, JICA has dispatched 3 long term experts and 14 short term experts apart from the Chief Advisor and Coordinator to work in the project and trained 25 counterparts in Japan in specific training/ Group Training Programme / Study- Visit. Under the training programme, 734 staff were trained under pre-cocoon, 436 under seed technology, 55 were trained under reeleing.

So also 3543 farmers, 200 seed rearers and 139 reelers were also trained.

The first years of the JICA project (August 2002 to July 2003) was called as period for preparation that was for strengthening of the identified BSFs, Grainages, TSCs, CRCs, Reeling TSCs, farmers and reelers to take up Bivlotine silkworm rearing and reeling under III phase of the JICA project. During this period 5 pre-trials were conducted with 199530 dfls. An average yield of 63.32 kg/ 100 dfls were obtained with average price of Rs.140.00.

The JICA trials were initiated from July 2003 and so far 10 trials were undertaken with the identified farmers from July 2003 to December 2005. A total 21.35 lakh dfls were incubated and supplied by NSSP and chawki reared in the identified CRCs. The chawki worms were distributed to the identified farmers. An average yield of 68.51 kg /100 dfls were obtained with average price of Rs.160.00. To create awareness among farmers for promotion and popularisation of bivoltine sericulture technology, extension activities like enlightenment programmes, group discussions and study tours were organised under the project.

During the year under report, 19 enlightenment programmes and 34 group discussions were organised. Apart from that, under the project, manuals, books and charts were prepared in local languages namely Kannada, Telugu and Tamil to create awareness & educate the extension workers and famers/ reelers.

ii) SERI - 2000

The agreement entered into with Swiss Agency for Development and Cooperation (SDC), for a project entitled SERI-2000 (1997-2004) at an outlay of 12.5 million

Swiss Francs was terminated prematurely by the Govt. of India due to irreconcilable differences. As on March, 2003, 43 project proposals under private sector at an outlay of Rs.1290.50 lakh had been approved. Under public sector, 30 projects at an outlay of Rs. 1312.33 lakh had been approved. These projects were continued till 31.7.2003. As on March, 2003, an amount of Rs. 19.09 crores has been incurred on both the Public and Private sector projects.

iii) Chhatisgarh Sericulture Project

The Govt. of Chhatisgarh is implementing Chhatisgarh Sericulture Project with financial assistance from Japanese Bank for International Cooperation (JBIC). The project is proposed to be implemented in two phases - the first phase of the project for development of tasar culture over a period of 7 years and the second phase for development of mulberry sericulture over a period of further 5 years proposed to commence from the 6th year of the first phase of the project. Thus the project would be implemented over a period of 10 years at a total cost of Rs. 748.55 crores.

The state share of Rs. 52.29 crores against the total project cost of Rs. 117.16 crores for first phase. iv) Manipur Sericulture Project Manipur sericulture Project is being implemented by the State Govt of Manipur with financial assistance from JBIC. The Project is to be implemented in 2 phases-first phase for mulberry silk and the second phase for eri silk. However, the loan agreement has been signed for implementation of only the first phase; the second phase is dependent on the progress of the first phase. Object of the first phase is to alleviate the poverty of the rural populations through generating the self-employment opportunities in the area of sericulture, and to meet the increasing demand for silk through increasing the silk production. The Revised

Action Plan of the first phase of the project has been approved in March, 2005.

Convergence Projects

CSB has adopted a convergence strategy to leverage funds and organisational capabilities for development in seri culture. By far the most successful has been the convergence with the programmes of the Ministry of Rural Development for creating sustainable infrastructure & livelihoods for persons below poverty line under the Special SGSY projects. The CSB lends its technical support & assistance to the States to draw up viable projects & to implement them, and in certain projects also pitches in with financial support.

Projects are being implemented with SGSY funds for development of all types of silks mainly non-mulberry silks i.e. muga, eri and tasar in the states of Assam, Arunachal Pradesh, Nagaland, Misoram, Bihar and Jharkhand. The projects adopt a soil-to-market approach, and aim to increase host plant development, quality egg production and supply, training & skill Upgradation, technological support in pre-cocoon & post cocoon process as well as market promotion with product design, development, diversification & product planning etc.

The Projects in hand in 2004-05 are:

i) Special SGSY project for development of Muga silk in Assam

(ii) Special SGSY project for development of Eri silk in Assam

iii) Special SGSY project in Goalpara, Kamrup and Sonitpur districts of Assam

iv) Special SGSY project for development of Muga silk in Arunachal Pradesh

(v) Special SGSY project for development of Eri silk in Arunachal Pradesh

(vi) Special SGSY project for development of eri silk in Nagaland

vii) Special SGSY projects for development of mulberry sericulture in the state of Mizoram

viii) Special SGSY projects for development of Tasar and Eri culture in Bihar and Jharkhand With this, a sum of Rs. 124.32 crores including Central funding of Rs. 51.36 crores has been attracted to the sericulture sector over & above the X plan outlays. This is expected to generate 971 M.T. of silk and employment for 715303 people in sustainable activities in sericulture & silk industry.

Other Projects:

1. Expansion of Sericulture in Non-traditional areas

For exploiting the vast potential available for development of sericulture in the non-traditional areas, CSB launched projects through the State Governments with financial & technical assistance under Catalytic Development Programme as below:

(i) Project for development of ericulture in Andhra Pradesh

(ii) Project for development of Oak Tasar in Uttranchal

(iii) Sonali Dhara Project in Assam for assisting commercial cocoon production.

(iv) Introduction of Ericulture in Punjab

(v) Introduction of Tropical Tasar, Muga & Eri culture in Uttranchal

(vi) Introduction of Ericulture in Chhatisgarh

(vii) R&D project on Eri culture in Tamil Nadu, Pondicherry and Kerala Introduction of Eri culture in Uttar Pradesh

(viii) Micro project for development of Oak Tasar Culture in Manipur Total cost of all the projects is Rs. 6.88 crores which includes Central share of Rs. 3.34 crores. The output is envisaged at 50 M.T.

2. Projects for Post Cocoon Technology Induction

CSB continued its emphasis on increase of productivity and quality of silk in the post cocoon sector through upgradation of technology & development and commercialisation of low cost user friendly devices for reeling & spinning, cocoon drying etc. which have helped to reduce drudging and increase efficiency, productivity and quality of produce. CSB also stepped up its financial and technical support for adoption and popularisation of these technologies. The main projects included:-

i) Establishment of Tasar Production-cum-Training Centre

Two centers to provide training to poor weavers in advanced technologies to enable them to produce quality tasar products were set up in Dhalbhumgarh block of east Singhbhum District.

ii) Vanya silk Production-cum-Training Centres (PCTCs)

Four centers, one each in Jharkhand, Chattisgarh & two in Assam, were established as field level units of the Product development Cell of the Central Silk Technological Research Institute (CSTRI), Central Silk Board, Bangalore to introduce new designs, products and technologies in the non-mulberry post cocoon sector in a cost effective manner. Rs.21 lakhs was the releases made by CSB to assist the establishment of VPCTC.

iii) Vanya Silk Mart

To strengthen the demand pull for Vanya silk products, Central Silk Board has hired 273.30 sft shop area in the Handloom Haveli, Bikaji Cama Place, New Delhi from the National Handloom Development Corporation (NHDC). The NGOs who were part of the UNDP assisted project would be given priority to showcase and sell the products.

iv) Vanya Web Sites

Central Silk Board has registered web sites titled 'Vanyasilkmart.com' and 'Vanyasilkgifts.com' aimed at cataloguing the product range of the small unorganised producers of Vanya silks in the Country. The web sites being developed would in future serve as a window to the wonderful world of Vanya silks and would also provide e-shopping facilities.

3. Development Projects in NER

The thrust on development of North East Region was continued. Central Silk Board implemented Central and centrally sponsored schemes in the North-Eastern region for development of Sericulture and silk industry in the state with cost sharing by the centre and state in the ratio of 90:10 respectively. Under the Centrally sponsored Catalytic Development Progrmame of the X plan, there is thrust inter-alia on development of non-mulberry sector. In 2003-04, 24.27% of the Sericulture budget amounting to Rs.22.50 crores was earmarked & spent on the NER. In 2004-05, the earmarking is 24.96%, amounting to Rs. 30.00 crores. Besides under the SGSY project of the Ministry of

Rural Development also financial assistance is being provided for various sericulture projects being implemented in the states of Assam, Arunachal Pradesh, Mizoram & Nagaland with CSB's technical assistance. With this convergence strategy, as described above, a substantially higher availability of funds was offered for the sector.

Central Silk Board

The Central Silk Board is a statutory body under the administrative control of the Ministry of Textiles, Govt. of India, constituted by an Act of Parliament (Act NO.LXI of 1948), The Board has a composition of 36 members including the Chairman, Vice Chairman, Member Secretary, representatives from the Lok Sabha and the Rajya Sabha, nominees of the Central and State Govts. and representatives from Industry. The functions of the Central Silk Board are mainly R & D, and Training, Basic Seed Support to States, Technology seeding and offer of supplementary developmental assistance to States. The Board has a work force of 4523 employees (including scientists) and about 2786 farm workers all over the country.

Under the Constitution, Silk Industry would fall under Entry 52,List 1 (VII Schedule), while trade, commerce and the production, supply and distribution of products of the Industry fall in Entry 33 of List III. Most States have their own departments to look after development of sericulture.

Support network of Central Silk Board

a) Research and Development

The Central Silk Board has three main research institutes

located at Mysore (Karnataka), Berhampore (West Bengal) and Pampore (J&K) and two non-mulberry research institutes at Ranchi (Jharkhand) for Tasar and at Lahdoigarh (Assam) for Eri and Muga. All these institutes are supported with Regional Research Stations and Research Extension Centres. Central Silk Technological Research Institute, Bangalore deals with post cocoon technology sector, supported by a network of Silk Conditioning and Testing Houses, Demonstration Cum Technical Service Centers and Eco Testing Laboratories. It performs in the sphere of silk the load of responsibilities discharged for other commodities by the Textiles Research Organisations. To offer support to the main research institutes and industry, a few specialised laboratories like Silkworm Seed Technology Laboratory and Seri-Biotechnology Research Laboratory both at Bangalore(Karnataka) and Central Sericultural Germplasm Resources Centre, Hosur (Tamil Nadu) is functioning.

b) Training

The CSB runs structured courses in its main institutes and imparts training in the post cocoon area through its Mulberry Demonstration cum Training Centres and Reeling Spinning Training cum Demonstration Centres.

Basic and Commercial Seed

i) Mulberry

The CSB has a chain of Basic Seed Farms supplying basic seeds to the States. Its Commercial seed production centres augment the efforts of the States in supplying commercial silkworm seed to farmers. CSB's commercial seed production involvement is about 8 per cent of the country's current estimated absorption.

ii) Non-mulberry

CSB has established a Basic Tasar Silkworm Seed Organisation (BTSSO) at Bilaspur to organise the production of tasar nucleus and basic seed. CSB has a Central Tasar Silkworm Seed Station at Lakha which supplies nucleus seed to 21 Basic Seed Multiplication and Training Centres (BSMTCs) for further multiplication and supply by the seed multiplication centres owned by the States. To supply muga and eri silkworm seed, the CSB has established 7 Muga Basic Seed Farms, 2 Muga SSPCs and one Eri SSPC in North Eastern region.

Price Support for Tasar and Muga

CSB operates one Raw Material Bank (RMB) for tasar at Chaibasa in Bihar with sub-depots at Raigarh and Bhagalpur and a Muga Raw Material Bank (MRMB) at Sibsagar to provide fair price support to the primary products under the price stabilisation scheme,

Pre-shipment (export) inspection

The CSB conducts voluntary inspection of silk goods meant for export at its 8 Certification Centres located at New Delhi, Mumbai, Kolkata, Bangalore, Chennai, Srinagar, Varanasi and Bhagalpur. 6 of these Certification Centres are attached to the Regional Offices located at the above places.

Apart from these, Regional Development Offices are functioning at Malda, Guwahati, Bhubaneshwar, Lucknow and Patna. From 1st May, 2001, Central Silk Board has started levying inspection fee under voluntary quality inspection scheme. e) Post Cocoon services to industry: CSB has 6 Silk Conditioning and Testing Houses (SCTH) (Bangalore, Dharmavaram, Kanchipuram, Jammu, Srinagar and Malda and has set up 4 Eco Testing

Laboratories (Bangalore, Varanasi, Jammu and Bhagalpur). Besides this, 16 Demonstration cum Technical Service Centres (DCTSC), 3 in West Bengal, 2 each in UP and Orissa and one each in Maharashtra, Bihar, Madhya Pradesh, Tamil Nadu, Andhra Pradesh, Jammu & Kashmir, Kerala, Karnataka and Assam, are functioning.

Indian Silk Export Promotion Council

The Indian Silk Export Promotion Council (ISEPC) incorporated on 16th June, 1983 as a company under Section 25 of the Companies Act, has its headquarters at Mumbai, and a membership comprising exporters of silk. It is entrusted with the responsibility of export promotion activities in respect of natural silk goods of the country. The affairs of the Council are managed by a committee of Administration (COA), comprising of a Chairman and 26 other members including 6 Government nominees.

The Council is eligible for admissible grant-in-aid from the Market Development Assistance (MDA) Fund of the Government of India for approved Code activities. The export promotion activities of the Council are participation in international fairs, organising generic promotion stalls in domestic fairs like TEXSTYLES India, publicity in foreign trade magazines, sponsoring of sale-cum-study tours and buyer-seller meets, dissemination of overseas trade information among exporters etc. The ISEPC is a small body, with hardly 1835 members. Its MDA grant is only about Rs. 56.63 lakhs per annum (2004-05).

10

Commercial Production of Silk

Silkworms

Silk is one of the oldest known textile fibers and, according to Chinese tradition, was used as long ago as the 27th century BC. The silkworm moth was originally a native of China, and for about 30 centuries the gathering and weaving of silk was a secret process, known only to the Chinese. Tradition credits Hsi-ling-shi, the 14-year-old bride of the Emperor Huang Ti, with the discovery of the potential of the cocoon and the invention of the first silk reel. China successfully guarded the secret until AD300, when Japan, and later India, penetrated the secrecy.

Silkworm, common name for the silk-producing larvae of any of several species of moths. Silkworms possess a pair of specially modified salivary glands called silk glands, or sericteries, which are used in the production of cocoons.

The silk glands secrete a clear, viscous fluid that is forced through openings, called spinnerets, on the mouthparts of the larva; the fluid hardens as it comes into contact with air. The diameter of the spinneret determines the thickness of the silk thread produced.

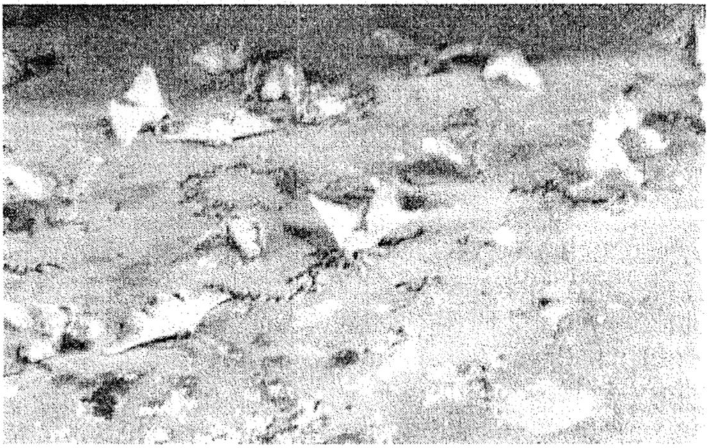

The silkworm moth

The best-known silkworm is the larvae of the common, domesticated silkworm moth. This moth, native to China, was introduced into Europe and western Asia in the 6th century AD and into North America in the 18th century. The moth has been cultivated for many centuries and is no longer known in the wild state. Breeders have produced many varieties of the moth, the most important of which produce three broods of young annually.

Gathered cocoons

A typical adult silkworm moth is yellow or yellowish-white, with a thick, hairy body, and has a wingspread of about 3.8 cm (about 1.5 in). The adult has rudimentary mouthparts and does not eat during the short period of its mature existence; the female dies almost immediately after depositing the eggs, and the male lives only a short time thereafter.

The female deposits 300 to 400 bluish eggs at a time; the eggs are fastened to a flat surface by a gummy substance secreted by the female. The larvae, which hatch in about ten days, are about 0.6 cm (about 0.25 in) long. The larvae feed on leaves of white mulberry, Osage orange, or lettuce. Silkworm caterpillars that are fed mulberry leaves produce the finest quality silk. Mature larvae are about 7.5 cm (about 3 in) long and yellowish-gray or dark gray in color.

Baskets of silk cocoons on the way to the factory

About six weeks after hatching, the common silkworm stops eating and spins its cocoon. The length of the individual fiber composing the cocoon varies from 300 to 900 m (1000 to 3000 ft). The silkworm pupates for about two weeks; if allowed to complete its pupation period, it emerges as an adult moth. Tearing during emergence damages the silken cocoon beyond commercial use. Therefore, in the commercial production of silk, only enough adult moths are allowed to emerge to ensure continuation of the species. Most of the silkworms are killed by heat, either by immersion in boiling water or by drying in ovens.

Sorting the silk cocoons

Sericulture, or the raising of silkworms, involves the incubation of the tiny eggs of the silkworm moth until they hatch and become worms. After hatching, the worms are placed under a layer of gauze, on which is spread a layer of finely chopped mulberry leaves.

For six weeks, the worms eat almost continuously. At the end of this period, they are ready to spin their cocoons, and branches of trees or shrubs are placed in their rearing houses. The worms climb these branches and make their cocoons in one continuous thread, taking about eight days for the process. The amount of usable silk in each cocoon is small, and about 5500 silkworms are required to produce 1 kg (2.2 lb) of raw silk.

After the complete cocoons have been gathered, the initial step in silk manufacture is to kill the insects inside them. Thus, the cocoons are first boiled or treated in ovens, killing the insects by heat. The silk fiber is obtained from the cocoons by a delicate process known as reeling, or filature. The cocoons are first heated in boiling water to dissolve the gummy substance that holds the cocoon filament in place.

After this heating, the filaments from four to eight cocoons are joined and twisted and are then combined with a number of other similarly twisted filaments to make a thread that is wound on a reel. When each cocoon is unwound, it is replaced with another cocoon. The resulting thread, called raw silk, consists usually of 48 individual silk fibers.

Ready for reeling

The thread is continuous and, unlike the threads spun from other natural fibers such as cotton and wool, is made up of extremely long fibers. Along with cocoons damaged by emerging worms used for breeding stock, the filaments from the coarse outer portion of the cocoon, which is removed by brushing before reeling, and the inner portion of the cocoon, which remains after reeling the raw silk, are mixed to produce a low grade of silk staple that is spun into yarn.

Unwinding the silk filament from the cocoon

The next step in the processing of silk is the twisting of one or more threads of the raw silk into a strand sufficiently strong for weaving or knitting. This procedure is called throwing. Four different types of silk thread may thus be produced: organsine, crepe, tram, and thrown singles.

Organsine is a thread made by giving the raw-silk thread a preliminary twist in one direction and then twisting two of these threads together in the opposite direction at the rate of about 4 turns/cm (10 turns/in).

Removing strands of silk from cocoons and feeds them onto a reel

Crepe is similar to organsine but is twisted to a much greater extent, usually between 16 and 32 turns/cm (40 and 80 turns/in). Tram is made by twisting in only one direction two or more raw-silk threads, with 8 to 12 turns/cm (20 to 30 turns/in). Thrown singles are individual raw-silk threads that are twisted in only one direction, the number of turns depending on the quality of thread desired.

In general, organsine thread is used for the warp threads of materials, and tram threads for the weft, or filling. Crepe thread is employed in the weaving of characteristic crinkly fabrics, and single thread is used for sheer fabrics.

11

Biomass Gasification

Mulberry silk production, which accounts for about 90 per cent of Indian silkworm breeding (sericulture), is concentrated in the States of Andhra Pradesh, Karnataka and Tamil Nadu in the south of the country. States towards the northeast produce such non-mulberry silks as tassar, muga and eri. Sericulture has spread over the years and now provides employment for more than 5 million people living in 59,000 villages and covers an area of 282,000 hectares.

An efficiency survey of 236 cooking ovens found that charkhas consume between 2.44 and 3.20 kilograms of wood per kilogram of cocoons processed, while cottage basin ovens consume 1.5 to 2.0 kilograms. It was then calculated that charkhas need about 1,300 kilocalories of energy to process one kilogram of cocoons and that cottage basins require about 875 kilocalories. This difference in energy needs can be explained by the longer operating periods and larger cooking vessel areas of charkhas.

The main drawbacks of the traditional ovens are that they do not allow fuel consumption to be controlled and that large fluctuations occur in such process parameters as water level and temperature. There was plenty of scope, therefore, for improving the efficiency and design of the ovens. As a first step, existing ovens were retrofitted to control their burning rates, maximise their flue gas heat recovery and reduce other losses. Retrofitted ovens allowed modest energy savings of about 25 per cent, but this was not enough to encourage silk reelers to face the inconvenience and expense of retrofitting their ovens. To be economically viable, the project had to come up with an alternative design that was capable of meeting the energy needs of silk-reeling units while generating substantial fuel savings.

Faced with the challenges of evaporation losses and the need for controlled burning, a gasifier system appeared to be the best option, and TERI started to develop a suitable gasifier-based silkreeling oven. A downdraft, throatless gasifier was selected because less tar and particulate matter are produced in its raw producer gas and it allows the use of relatively large wood chips and other fuels compared with updraft and cross-draft gasifiers. A water seal tank is placed under the gasifier to collect ash and cinders.

The main goal of the project was to improve the productivity and profitability of post-cocoon processing in the silk industry, including reeling and dyeing. Project activities fell into five main categories: technology and system development, economic assessment, market development, entrepreneur development, and human resource development. The final product—the SERI-2000 oven—evolved from a process of prototype development and testing, which included inputs from silk experts, engineering consultants and manufacturers.

The oven is now commercially marketed by two licensed manufacturers. The entire process, from laboratory prototype to final commercial model, was completed in just over three years. Before the ovens were put on the market, they were extensively field-tested to quantify their advantages over traditional ovens. An economic assessment showed that the new technology had a payback period of less than a year in terms of reduced fuel consumption and increased productivity and profitability. In fact, the new ovens allow fuel savings of 70 per cent or more, representing nearly 300,000 rupees a year.

Development of the oven involved the design of a series of laboratory prototypes, each of which was field tested and adapted to create the subsequent improved prototype. Data regarding quality and yield improvements were collected, and reelers and silk experts involved in the testing provided feedback and comments at every stage. Comparative data for both the traditional and the new ovens—including cocoons processed, fuel consumed and silk produced—were collected.

With the gasifier ovens, fuel consumption declined by 57.3 per cent, water consumption by 28.4 per cent and renditta (the number of kilograms of cocoons required to produce one kilogram of silk) fell by 3.6 per cent. Quality improvements included fewer breaks during winding and a cleaner, neater, stronger product. The project team also carried out studies to establish the most suitable capacity for each of the applications of the gasifier.

Manufacturers and local marketing agents helped to test commercialisation of the new technology through a direct marketing approach. To start with, a promotional test-marketing phase was initiated to generate awareness of the benefits of the new systems among potential users. Brochures in local languages were distributed to silk reelers, and demonstrations were held at cocoon markets.

A team of consultants and project representatives then assessed the technical, financial and marketing capabilities of about 15 manufacturers to find the most appropriate partners for the commercial distribution of the ovens.

The two manufacturers chosen were 2M Industries of Bombay and Silktex Industries of Kanakpura, and the technology was transferred to them through a license agreement. Members of the project team trained the manufacturers in fabrication details and quality-control procedures, and stringent quality checks were carried out. Discussions were held with financial institutions so that a suitable marketing scheme could be worked out. One of these financial institutions, SDC, designed a financial package to facilitate the first stage of commercial distribution through awarding favourable credit facilities to the buyers of the first 100 gasifier-based systems.

During the system development, fieldtesting and marketing phases of the project, the team entered into partnership with the following individuals and public- and private-sector institutions and organisations:

— private manufacturers - 2M Industries of Bombay and Silktex Industries of Kanakapura;

— silk experts - D. Mahadevappa and T.S. Nagaraja;

— design consultants - the Industrial Design Center, IIT of Bombay and Kvaerner Powergas of Bombay;

— numerous potential users of the new technology;

— international experts - Sorane SA (Switzerland) and Ashton Court Consultants (United Kingdom);

— the donor - the Swiss Agency for Development and Cooperation; and

— Departments of Sericulture in Andhra Pradesh, Karnataka and Tamil Nadu.

The successful development and demonstration of a gasifier-based system for thermal applications in the silk industry has had spin-off effects in many other industries that need to generate heat. The new technology is already being used, among other activities, for rubber drying, textile dyeing and large-scale cooking, resulting in reduced fuel costs and improved product quality because of better temperature control.

The system could also be replicated in many other small and medium-sized industries throughout India, including: puffed rice manufacture, tapioca making, baking, the hotel trade, lead recovery from used batteries, wire enameling, tobacco curing, lime kilns, mini cement plants, gur making, tea drying, brick drying, khoya making, small-scale cardamom production, coffee curing, food processing, carbon dioxide manufacture and copra drying. Other countries with similar conditions and large numbers of similar industries could also benefit, including Bangladesh, Malaysia, Myanmar, Sri Lanka and Thailand.

The main objective of this project was to develop a commercial product based on gasifier technology for use in the silkreeling industry. Its success was the result of designing a technology that was based on research into the processes involved and constant feedback from field-level interaction at all stages. The development and adaptation of renewable energy devices for small-scale rural enterprises are both time-consuming and expensive. Over the years, the Indian silk industry has become highly dependent on government policies and interventions.

Most of the technology upgrading and improvement in the sector has been driven by government support and there has been very limited private-sector involvement. Machinery and equipment have also been marketed through government subsidies or incentives, and there have been no direct links between manufacturers and silk reelers. The project was innovative, therefore, in that it put manufacturers and their local agents into direct contact with reelers. This meant that the benefits of the new technology had to be made very clear to its potential users, who were used to the security of government financial support.

Reelers expected to be able to buy the ovens with subsidies or through credit schemes, but they have a poor repayment record so bank loans did not seem to be a viable option. To overcome this challenge, favourable credit arrangements were used to finance the distribution of a few gasifierbased systems within the silk-reeling community as a way of demonstrating their economic and other advantages. The idea was that, rather than using financial incentives to create artificial markets, potential users should be encouraged to adopt new technologies because of their inherent economic viability and the clear benefits of using them.

In this way, renewable energy projects can become truly successful and sustainable because the people involved in them have a clear interest in keeping them going. Small and micro enterprises contribute about 7 per cent of the gross domestic product (GDP) of India. In recent years, energy prices have been increasing rapidly. Diesel oil prices, for example, increased from 7.95 rupees per litre in January 1997 to 16.59 rupees per litre in August 2002. These price increases hit small-scale producers particularly hard.

Biomass fuels can replace fossil fuels in several small industries that use large amounts of energy for thermal applications. There are immediate plans to introduce the new technology into other applications, such as textile dyeing, rubber drying, hot water and steam generation, and non-ferrous metal treatments. However, this will not be a simple add-on process because much of the existing process equipment used by the small-scale industries will have to be adapted to accommodate the gasifiers. For this, the project will collaborate with specialised engineering experts from the various industries involved and will carry out extensive field-testing and demonstrations. Once suitable technology packages have been developed, the project team will turn its attention to the financing and marketing aspects.

12

Project Cycle Analysis for Sericulture

Silkworms are stenophagous insects that are fed solely mulberry leaves. They through four different stages, viz., egg, larva, pupa and moth for the complete the life cycle. Adult moths have creamy white wings with brownish patterns across the front wings. The body is very hairy and the wingspan is about 50 mm. Adult females are larger and less active than males. Male moths actively crawl around looking for females. They will copulate for several hours.

After hatching from eggs, the silkworm larvae grow by feeding on mulberry leaves, except during moulting, during immature stage they moult four times then become matured larvae in about 25 days. During this period, the silk glands utilise proteins, amino acids, carbohydrates, etc., present in mulberry leaves as raw material and synthesise fluid silk which is secreted into the gland sacs.

The fluid silk, thus secreted, comes out as a silk filament through the spinneret. The silk filament is a continuous thread of great strength measuring from 500 - 1500 meters in length. Single filaments are too thin for utilisation. For production purposes, several filaments are combined with a slight twist into one strand. This process is known as "silk reeling or filature"

Silk production encompasses planting mulberry trees up to production of eggs from moths, from supervision of single instar of silkworms until the cocooning, the harvesting of cocoons and processing to thread, yarn and fabrics. The complexity of the production activities needs manual labour force. Sericulture activities are broadly classified into two: the agro-based sector and the industrial sector.

The agro-based part involves two distinct phases of activities i.e. mulberry cultivation and silkworm rearing. Silkworm rearing is again differentiated into two: young age rearing from 1st instar to 2nd instar. The 3rd instar although an intermediate stage, is considered as young age. Rearing of 4th and 5th instars comes under late age rearing.

Mulberry Trees

In flora categories, mulberry belongs to the family Moraceae, genus Morus. The genus Morus has many species and subspecies. Again, under the species and subspecies, there are several varieties. The natural distribution of the genus Morus covers western Africa, Middle Eastern Asia and Central and South America. It is not a natural plant of Europe. The mulberry tree is a quickly growing tree providing a lot of leaves. It grows in subtropics and temperate latitudes in evergreen tropics and reaches a height of up to 20 to 25 metres.

Mulberry leaves are the only foods tuft accepted by the silkworm. Therefore every single silkworm breeding activity that is planned has to make sure that a sufficient amount of mulberry trees is grown for food supply. On the basis of usage of mulberry, the mulberry field is classified into the type utilised for rearing young silkworms and the type utilised for rearing older ones. Regarding the mulberry field for rearing young silkworms, some areas are used exclusively for spring rearing and other areas are used exclusively for summer and autumn rearing.

The type of mulberry field is selected on the basis of weather conditions of each region, soil conditions and coordination with silkworm rearing.

Production of Egg

The production of silkworm eggs correlates closely with breeding success and is mainly carried out in specific institutions. Silkworm eggs are broadly classified into parent eggs and hybrid eggs. In the case of the parent ones, there are two types, parent eggs which are used for preparing hybrid eggs, and grandparents eggs which are used to produced the parent eggs. The hybrid eggs are known as the commercial silkworm eggs from which the larvae are used for producing reeling cocoons from which no adult moths are allowed to emerge. All other eggs are used for producing breeding adult moths. Silkworms have a wide range of races, some of which are classified in the following manner:

— According to their native region: Chinese race, Japanese race, European race, tropical race, etc.

— According to voltinism: univoltine, bivoltine and polyvoltine or multivoltine.

— According to moulting feature: trimoulter, tetramoulter, pentamoulter and rarely hexamoulter and bimoulter.

Silkworms are also classified according to colour, shape and size of eggs, larvae, pupa, moth, cocoon, as well as on the basis of larval pattern, strength, course of larval period, quantity of fibre, reelability, etc. According to their utilisation, silkworms are classified, besides the race 36 used for producing silk fibres there are some which are used for high quality fabrics (fine and lousiness-free) and for special textures.

Breeding

Immediately after hatching, the worms start eating mulberry leaves. A hatched silkworm larvae weighs approximately 0.45 mg and exclusively eats leaves of mulberry tree (Morus spp). The larvae develop in four stages for about 25 days, and between each of the four stages they undergo moulting and at the end they spin cocoons.

A silkworm is ready to spin a cocoon when it weigh's about 5 g. First, a thread capsule is spun and after that the worm spins itself inside its cocoon. A cocoon weights about 1.25 - 2.50 g and has a filament of 500 - 3000 m long, dependent on silkworm breed and cultivation season. The cocoon spinning is completed in two days. In the next two-three days, the larvae inside the cocoons moult and become pupa. The pupa takes 10 days to transform into a moth which is white-grey coloured; The female moth produces 300 - 500 eggs.

The quantity of mulberry leaf supplied during the rearing of advanced stage larvae amounts to 90% of the total supply of mulberry (800 kg - 1000 kg) during the entire larval period. During the fourth and fifth instar the quantity of mulberry feeding is very large and consequently the labour requirement also becomes high. Therefore silkworm rearing from the fourth instar until fresh cocoon spun is commonly accomplished by silk farmers.

Silk Yarn Production

The silk yarn production begins with selecting for their quality. The cocoons will be sorted into normal and abnormal cocoons. The fresh cocoons normal are reelable to produce raw silk but the abnormal cocoons are unreelable. The abnormal cocoons are double cocoons, perforated cocoons, internally and externally soiled cocoons, thin-end cocoons, thin-middle cocoons, malformed cocoons, etc.

After the first selection of cocoons, they have to be dried. The first goal of cocoon drying is the protection of cocoon quality, to preserve conditions for reeling cocoons and prevent damage that might be caused by long periods of storage. Drying kills the pupa and evaporates moisture that would otherwise ruin cocoons. Dried cocoon storage conditions are designed to keep the raw material for long periods without any damage from moulds and pests.

Cocoon storage should preferably be built with double walls. Cloth or polyethylene bags are recommended as containers for dried cocoon storage. The technological process for silk reeling is shown in figure 1.

In most modern factories which aim at producing high-grade raw silk, the cocoons are graded on visual inspection or by mechanical tests are actually mixed as in the required proportions. This is called cocoon mixing or blending and is done to ensure speed and uniformity of reeling as well as to obtain the desired effect in raw silk. The reelable cocoons have then to undergo the reeling processing, after which they are either processed further

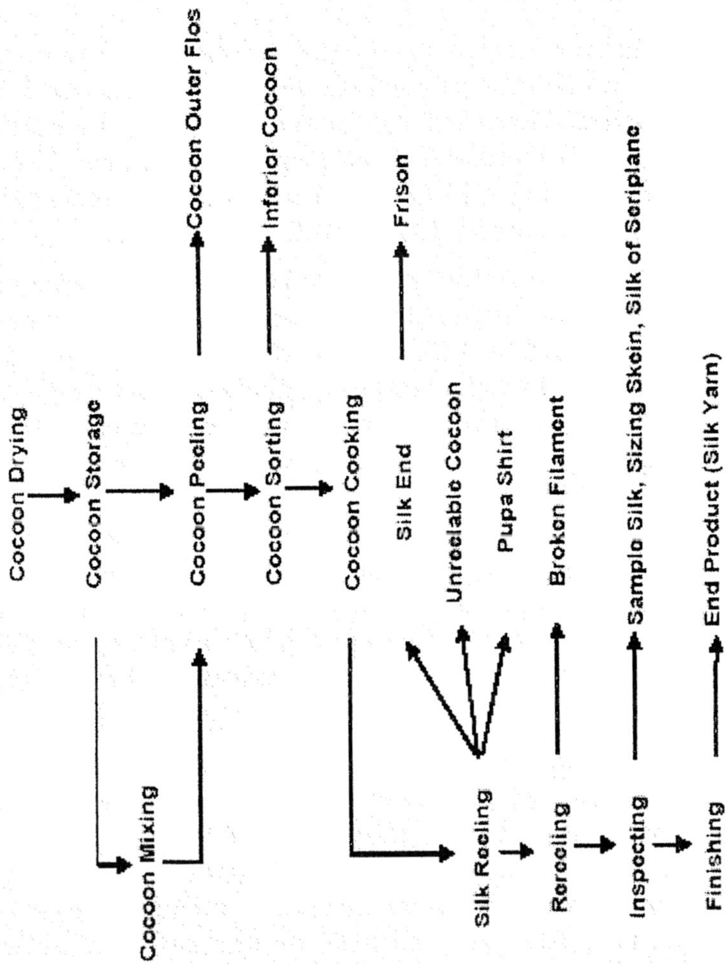

Figure 4. Technological process for silk reeling.

to raw silk or to thrown silk. The unreelable silk is processed further to spun silk.

Origin of Silkworms

The origin of silkworms is not definitely known. In general the Chinese Himalayan region is understood to be the origin. There are also opinions naming the north-eastern Indian Himalaya as an origin of silkworm. There is evidence using C14-Isotope-Test that silk production was occurring around 2850-2750 B. C. in China.

In China, sericulture seems to have had its beginning in Huango and Yangtze (Chang Jiang provinces). Diverse literature sources tell of the development of silk production (silkworm breeding, planting and usage of mulberry trees, spinning of yarn and weaving of fabrics) reaching back to emperor Huang-ti and his wife Hsi-ling-shi around 2600 B. C.—eventually even further back to 5000 B.C.. Others tell of exploration in the 1122 - 255 B. C.. Sericulture extension was supported at the highest governmental level and sericulture spread out to small farmers and their education. After short time silk production had spread over approximately half of the former imperial China.

China clearly realised the value of its sericulture, not only because of the end product but due to the knowledge gained over thousands of years about silk production, moriculture and processing of silk. Referring to Silbermann a Chinese princess brought eggs of silkworms inside her headdress for the occasion of her marriage to Khotan in East-Turkestan (northwestern China). This led to a sericulture industry that lasted for hundreds of years. In the year 552 A. D. two Nestorian monks smuggled silkworm eggs to Byzantium, which was the first step for expansion in the western direction.

The know how of silkworm breeding and of mulberry cultivation came to Korea and Japan with Chinese immigrants in between the years 200 B. C. to 300 A. D.. A further expansion came with captures, conquests and population movements of Persians and Arabs. The Moors brought sericulture later in the 9th century to Sicily and Spain. From Spain, sericulture spread to northern Italy, the region that had most of the European important producer- and trade centres in the 15th century. Through annulment of the edict of Nantes 1685, the Huguenots were exiled. They took sericulture with then to Germany, Switzerland, the Netherlands and England.

Due to less favourable climate conditions, these countries created a highly developed silk industry. This was the time of separation between raw silk producing countries and those of silk industry, referring to Jakobi. An figure that still dominates the geographical structure of global silk production, 40 Alexander von Humboldt in 1809, tells of the importation of sericulture into Central- and South America by Europeans in the 16th century. However successful implementation of some experiments secured Mexico at this time.

Availability of silkworms and mulberries is related to climate and ecological conditions (rainfall, temperature, relative humidity, soil quality, etc.). According to altitude and longitude, sericulture is focused on regions between the 30th and the 35th northern latitude. These are humid to sub humid tropics. The large genetic variability concerning mulberries as well as silkworms opens up many possibilities for breeding and selection, e.g. the expansion to the Fore Asia to approx. 45th latitude degree.

Even more spectacular, sericulture reached production levels up to the 60th latitude degree, e.g. coastal regions in Norway and Sweden. The general technical know-how was mainly spread with imports of silkworms. The most important reason for expansion and changing denotation of sericulture is moreover of an economic nature. With increasing labour costs and few mechanisation possibilities the production of raw silk became unattractive. In Europe and Japan production practically disappeared completely.

In the USA breeding attempts could not keep pace with other countries. In the meantime labour diversification between production of cocoons and trade oriented commercial processing to fabrics. Therefore China, India, Turkmenistan, Uzbekistan and Kirgisistan and newly also Brazil remain, as main producer countries while main processing countries are the USA, Italy and France.

Table 1. Silk, raw and waste production

Silk, Raw and Waste	Year						
Production (Mt)	1961	1970	1980	1990	2000	2001	2002
Brazil	101	259	1,300	1,693	1,500	1,400	1,450
China	6,268	13,516	35,393	55,003	78,201	94,201	94,201
India	1,264	2,258	5,041	11,800	15,197	15,000	15,000
Italy	761	310	18	2	12	12	2
Japan	18,680	20,516	16,155	5,721	650	559	559
Vietnam	200	300	360	500	3,000	10,000	12,124
World	32,503	45,762	68,741	83,401	106,991	130,771	132,434

Worldwide silk imports reached a total value of US$ 448 million in 2001. Italy and Japan are the most important silk importing countries. The export oriented silk processing industry of Italy is completely dependent on imported raw silk material, of which Italy only buys the highest quality. In Italy an amount of 50 kg raw silk has a import value of US$ 1,275. The largest supplying country of raw silk for Italy is China, at it is for the global market.

The case of Japan is conspicuous with figures showing Japan as one of the most important global producers in 1961 (57% of world production) but decrease to 559 mt (0.004% of world production). Japanese production technology had been predominant in comparison to Chinese and Japanese international sale positions were even more superior. Since Japan's industrialisation and non-agricultural income possibilities increased, national silk production was not profitable anymore. India pushed into this market gap but China was even stronger than some other countries.

In consequence, total production figures grew four times from 1961 and 2002. Nevertheless, volume of world silk production is marginal in comparison to other textile fabrics. In future, production probably will be shifted from richer countries to poorer ones. Sericulture will be moving more and more into tropical regions. Sericulture is seen as a connoting economic approach of development for rural areas, well suited for diversification of agricultural production.

As a labour intensive agricultural production branch sericulture offers agricultural working places and income possibilities but also in preliminary and downstream industrial sectors, like reeling work, weaving mills, spinneries and dyeing factories. Additionally, the products of natural silk product can support diversification of agricultural export commodities. But introduction of sericulture with high hopes of easy and quick success are amiss.

Sericulture consists of strong time controls concerning the provision of inputs to follow phase production. Therefore not only production technology but also exact production programmes and dispensation systems for supply on schedule of breed have to be evolved for 42 implementation of sericulture projects. Also fresh cocoons are perishable goods and have to be either dried or reeled one week after cocooning. Implementation of silk production to export quality requires marketing measures to be created. These should be located together with cocoon-collection quality determination facilities and, if required, cocoon drying and cocoon storage places.

Sericulture also requires an intensive consulting service by professional personnel. Only through a tight connection between research institutes on regional level (e.g. in China on province level) and consulting services the ascertained technology transfer can occur. Research centres have to be built up with responsibility for breeding and development of adapted mulberry varieties and silkworms as well as to find solutions for production technical problems and the development of new technologies.

Demand for capital for sericulture is evaluated as dependent on production system intensity by different authors. Independent of the amount of investment required is the development of suitable credit programmes for small farmers and co-operation with credit institutions for financing is indeed important. Sharifi points out that silk production presents an advantage concerning loan repayment due to harvests already available in the first year of cultivation in comparison to many other permanent cultures. Thus, a more rapid start date of repayment and recirculation of loan money is possible through sericulture.

China is an example of integrated sericulture. Chinese silkworm breeding and the correlated mulberry cultivation produce a range. Moreover sericulture can be combined with tillage, animal- and fish farming systems. Increasing population densities made irrigated agriculture necessary for constant and higher yields.

Meanwhile implementation of fishponds (aquaculture) became possible. That way, a stable and highly productive farming system is created. One system with mulberry trees is to grow them on hillsides with irrigated rice terraces, irrigated rice in shallow areas and ponds whose fish production is secured through wastes from silkworm breeding. The manifold interactions allow a maximum usage of resources (land and labour) and a stability of farming systems. Stability or sustainability of farming system in addition to the production of silk is further advantage of sericulture in tropical regions.

Silk Production in Thailand

Archaeological discoveries in the village of Ban Chiang in the north east province of Udon Thani have led experts to believe that Thailand's sericulture history may be as old as China's. An extensive burial site at Ban Chiang yielded evidence of a complex civilisation involving silk production that dated back over 4000 years. In 1901, King Chulalongkorn made an attempt to upgrade the local silk industry by inviting a team of Japanese experts to aid production. In 1903, the Department of Silk Craftsmen was established under the directorship of Prince Phephatanapong. These early steps marked the beginning of rapid sericultural development in Thailand.

Mulberry trees were planted in the northeast, local silkworms were cross bred with the Japanese variety and spinning looms were replaced by more advanced ones.

Sericulture courses were taught throughout the Kingdom. By 1910, over 35 tons of silk were being exported annually Despite the intricate role Thai silk played in high society, a flood of imported fabrics including fabulous silks from China, Persia and Japan made it difficult for local silks to compete. Thai sericulture remained a small cottage industry, most active in the Northeast around Korat, until the mid 20th century.

Silk production gradually decreased over the following three decades due to a lack of government support and because of strong competition from foreign silks. An inability to improve antiquated techniques resulted in limited production. The FAO reported that a two year project to produce disease-free silkworm seeds/ hybrid eggs for sericulture farmers started in December 1997 with the additional objective to train staff of DoAE in sericulture.

A special training component for the rural disabled was inserted in December 1998. Eight disabled farmers learned the method of cocoon processing with the newly developed reeling machine as well as the making of silk products for the raw silk. The project's objective is to raise the net incomes from sericulture for approximately 8,000 farm households in the project area. The development and strengthening of farmer groups has been the foundation of all aspects of the project's fieldwork.

The project works closely with the permanent staff of the DoAE including three Sericulture Extension Centres. The farmer groups are assisted in all technical aspects of sericulture production from mulberry planting to 45 fabric design and weaving. The project has facilitated contacts and marketing agreements between the farmer groups and a number of private companies.

By February 2000 the project had supported 45 farmer groups covering approximately 200 villages and 8,000 farmer members. Monitoring data show that production levels and incomes have increased significantly as a result of project activities. In 2002 Thailand produced 1,510 mt silk.

Sericulture Development in Vietnam

Vietnamese sericulture began more than 3,000 years ago, probably under the influence of China, and has been practised ever since by most village families, largely for their own use and for the domestic market. Exporting started with the arrival of the French about 150 years ago and continued until the Second World War. In 1985, the Government launched the Silk Integrated Agro Project (SIAP) for the promotion of sericulture. The Vietnam Union of sericulture Enterprises (VISERI) was established in Bao Loc as the agency responsible for management, development and trade.

VISERI is part of the Ministry of Agriculture and Food Industry. Viseri proposed recently that Bao Loc is not the only silk production centre in Vietnam, because the climatic conditions in the Lam Dong highlands certainly make it one of the country's most suitable areas for sericulture. The subtropical climate (22°- 29° C) contributes to the availability of fresh mulberry leaves all year round. Elsewhere in the country, fresh leaves can not be obtained for about three months in the year because of heavy rains or extreme heat, and cropping can be carried out only 7 - 8 times yearly.

The mulberry plantation increased from 36,500 hectares in 1994 to 100,000 hectares in year 2000. Many areas in the country, which are now planted with coffee or tea, could easily be used to sericulture. A shift may become attractive to farmers since the income of Sericulturists can be as much as four times as high as those of coffee or tea growers. Sericulture is labour intensive and a one -hectare farm employs about 10 persons. Vietnam is expected to develop into an important international supplier of raw silk and silk yarn.

The Viet Nam Sericulture Corporation (Viseri) in conjunction with authorities in Lam Dong Province recently conducted an inspection tour of all silkworm farming households in Lam Ha District, the country's largest locality specialising in the growing of mulberry and the 46 breeding of silkworms. It was found that in May and June about 4,000 households had incurred VND216 million (US$14,000) in losses due to poor-quality silkworm eggs imported from China. This loss was largely due to degenerated species of silkworms and spoiled frozen eggs.

Chinese specialists in the inspection teams agreed that spoiled silkworm eggs are the main cause of damages to farmers and they vowed to compensate 60 per cent of losses to the farmers. The FAO reported that the silk production in Vietnam rose from 300 tons in 1970 to 12,124 tons in year 2002.

Sericulture Development in Malaysia

Malaysia's small sericulture industry mainly produces cocoons for export to Japan. The producer is Permit Suterasemai, a subsidiary of The Terengganu State Economic Development Corporation. The company's 540 ha have an output of 640 tons of mulberry leaves every month. Silk eggs are imported from several countries, including Japan. The country's imports of silk fabrics have risen rapidly in recent years, from 19 million USD in 1992 to 89.4 million USD in 1995. In the year 2000 the

United Nations Office for Project Services (UNOPS) reported a silkworm development project in Malaysia. The project has two major objectives:

- — To lay the foundation for the modernisation of the sericulture sector in the country through investment support to selected less-advanced co-operative farms and institutions providing sericulture support services.
- — To help raise incomes and improve working conditions in the selected co-operative farms, particularly for women who constitute the majority of sericulture workers.

To achieve these objectives, the project finances the following components: improvement and expansion of existing sericulture production; capacity building of sericulture support services; training and pilot activities and support for project management. This project will benefit about 2,900 members of sericulture work teams, equivalent to about 1,500 households or 6,100 people. It would indirectly benefit about 59,800 people additionally (co-operative members and their families) through increased incomes. Average incomes for workers in project co-operatives are expected to increase by up to 30% over 9 years. In 2001 Malaysia exported 29 mt reelable cocoons.

Sericulture Development in Indonesia

Attempts at developing sericulture in the Indonesian Archipelago date back to the 18th century. Zwaardecroon (1718-1725) and De Haan (1725 -1729), at that time Indonesia was a colony of Netherlander, Indonesia have produced silk yarn 34,5 lbs and exported 300 lbs to Netherlander in year 1735. The modern sericulture and silk processing industry in Indonesia started in 1950 when some Indonesian war veterans supported by ex Japanese war veterans initiated the sericulture industry as a home industry to support the existing traditional silk weavers with raw silk as their raw material to produce traditional silk fabrics.

The government appointed the Forestry Department to develop Mulberry Plantations and sericulture at the borders of forests, to protect the forest from people who destroyed forest for wire wood. By providing to forest workers, a barrier was created to protect the forests, especially on the island of Java. Cocoon production in Indonesia reached its highest level between 1962 and 1966, but collapsed in the 1970s due to the spread of the pebrine disease. From the 1990´s rising cocoon prices led to a production increase.

Development of silk production efforts are now concentrated on smallholder areas. According to data obtained from the Indonesian Natural Silk Community and the Sericulture Station in Bili-Bili, South Sulawesi, the farmers involved in sericulture production increased from 7,462.00 families in 1991 to 12,631.00 in 2001.

Most developing countries are raw material producers of agricultural commodities. Income generation in developing countries is therefore often dependent on agriculture in the form of small farming systems. Rural regions make about 70 - 90% of governmental areas in most of the world's developing countries.

In general, the rural population is sensitive to climatic or seasonal changes and thus, changes of market prices but moreover is strictly dependent on their resources (labour and land) as a source and insurance of income. In Indonesia income figures vary between rural and urban areas. In this case, income per capita should be differentiated into two levels of income, as the urban income is relatively higher than that of the rural population.

For the Indonesian rural population income generation is an important aspect. From the sustainable development point of view that this kind of development assistance is a great advantage for:

— cash-availability for educational purposes,

— low productivity sector agriculture,

— national problem of growing gap between poor rural and rich urban,

— diversification and increasing productivity of agriculture

Project Cycle for Sericulture in Central Java

The project idea is based on the possibility of generating additional income for rural households in Central Java, Indonesia through adopting silkworm/cocoons production. For simplifying the complexity of Indonesian households, six types of rural households are chosen as exemplary models. The first three are farmers in activities "without sericulture". The last three ones are farmers "with sericulture" in different land categories, starting from: 0.30 ha (small size of farm); 0.50-0.90 ha (middle size of farm); 1.00-3.00 ha (larger size of farm).

FAO explained that income-generating activities in the countryside can be adopted, if the rural dwellers can boost income and living standards through a number of rural activities: crop and livestock production, processing, para-. non-agricultural, artisanal and commercial activities. All these activities must fill full certain conditions: In addition to being technically feasible, they must also be economically and financially profitable. The income-generating activities should be profitable, that is to say they should produce substantial increase in income or surplus (one could also use the term profit) as a result of labor inputs. If we the term "substantial increase" (rather arbitrarily) as 50%, the project idea can be said to have passed its first test.

Besides a comparison of sericulture production with the current revenue in rural areas, sericulture should be compared with alternatives in crop production. Revenue from sericulture is higher than from cassava, maize or peanut. Sericulture shows higher economic profitability, however it may not be easy to adopt, because peanut, maize and especially cassava, do not require intensive daily labour.

The tropical climate in Central Java allows the mulberry to grow during the whole year so that sericulture production is practicle every month. Five sericulture period cycles were assumed in the first production year due to the growing phase of mulberry trees which takes six months. For the 2nd to 10th production year a number of ten period cycles were assumed. The project models were also calculated for ten years.

The mulberry section includes costs for land preparation, fertiliser, mulberry cuttings and mulberry planting, grass pruning and labour costs. The silkworm section includes the shed building and accessories for rearing.

Sensitivity analysis

Sensitivity analysis is the study of how the variation in the output of model (numerical or otherwise) can be apportioned, qualitatively or quantitatively, to different sources of variation, and of how the given model responds to upon the information fed into it Sensitivity analysis was created to deal simply with uncertainties in the input variables and model parameters. Sensitivity analysis measures the economic impact resulting from alternative values of uncertain variables that affect the economics of the project.

When computing measures of project worth for example, sensitivity analysis shows just how sensitive the economic pay-off is towards uncertain values of a critical input, such as the discount rate or project maintenance costs expected to be incurred over the project's study period. Sensitivity analysis reveals how profitable or unprofitable the project might be if input values to the analysis turn out to be different from what is assumed in a single-answer approach to measuring project worth. Sensitivity analyses have been carried out on four of variables: income variables.

The variables within the presented sensitivity analysis it was assumed at twenty percents of variation. Changes in production costs can take a great influence on income figures. The case of an increase of input costs required for sericulture will lead to a decreased income in the starting years. The challenge will be to determine and operationalise as to handle the single aspects that have influence on the costs.

On the other hand, a great impact is detected in the aspect of decrease of income. It has to be paid attention to production factors that have an impact on income figures of small land sizes of rural farm households joining the project. In the above calculations it was explored 57 that particularly the small ones need support in cases of income changes. Consequently, aspects that influence the income for those, thus, the small size sericultural harvests require a close and intensive consultation for farmers in their first years.

An environmental impact analysis report was completed by consultants in May 1997. The report concluded that no significant environmental concerns were present. Mulberry cultivation rather supports soil conservation and the pruning reduces unsustainable harvesting of fuel wood.

Project Plan

The project plan is evolved to present and coordinate a structure to ensure that stakeholders are consulted, and defines key decisions, information requirements and responsibilities at each phase in the life of a project. Required preparation, implementation and evaluation are feasible and accomplished. Drawing up the project plan evokes a differentiation of two participating groups.

Organisational Structure

The two differentiated aspects may be described as "nucleus" and "plasma". The name "nucleus" determines the non-governmental organisation Banyumas Sutera Alam and the "plasma" that means farmers of several regions in Central Java that surround Banyumas Sutera Alam.

The institution that is providing the Non Government Organisation with silkworm eggs is a breeding station run by the government. The supplement, reaching from delivery of eggs until giving young silkworms away to farmers is done by the Non Government Organisation. Farmers provide advanced silkworms, feeding the worms for two weeks and waiting for one week for processing of cocoons. The harvest of fresh cocoons is sold to the Non Government Organisation.

The Non Government Organisation takes on responsibilities for spreading of sericulture, silk production among rural smallholders and buys their produced fresh cocoons. Further processing of bought cocoons is then accomplished by the Non Government Organisation. Simultaneously, Non Government Organisation has also established market and technical assistance network with Indo Jado Sutera Pratama, Ltd. Co., a local silk thread factory. Indo Jado would buy the silk worm cocoons produced by the farmers and provide technical assistance to the farmers to improve their skill in silkworm farming to enable them to produce better cocoons. These products are sold on local markets to guarantee sustainability of the project.

Banyumas Sutera Alam is a new Non Government Organisation specialise in silkworm farming activity, in Kaliori-Village, Banyumas, Central Java, Indonesia. Banyumas Sutera Alam is part of Yayasan Sosial Bina Sejahtera (YSBS), founded 1972, is a non-profit organisation that belongs to Catholic Church diocese in Central Jawa. Banyumas Sutera Alam started the cultivation of silkworm farming in 1997 with a mission to help village people to have more prosperous life through silkworm farming. BSA has functions as a coordinator, young worm producer and purchaser of fresh cocoons. Activities are being planned as follows:

Preparation: An Experimental Project (EP) and on-farm demonstrations (Pilot Project /PP) are carried out before sericulture is promoted to potentially participating farmers. EP tests the sericulture production in project areas, including mulberry cultivation and silkworm production. On-farm demonstrations increase participation of farmers due to easily comprehensible training courses. The farmers can see and learn practically how sericulture production is executed and how it positively affects their income. Participation often increases when farmers have conceived thoroughly how to carry out sericulture.

The Experimental Project is planned for one year and a Pilot Project is prepared for two years. For mulberry cuttings required by participating farmers the Non Government Organisation has planted sufficient supply of mulberry trees. To cultivate an area of 1 ha mulberries an amount of 20,000 mulberry cuttings are needed. Mulberry cuttings can be cultivated from the branch of the mulberry tree itself. The Non Government Organisation cares for distribution and supply of mulberry cuttings from the Experimental Project so that farmers get cuttings on time to begin cultivating. After one-year growth farmers can transfer mulberry cuttings again to other farmers.

Training and extension service: Training has to be conducted after farmers display interest in joining sericulture activities. Training courses give practise oriented know-how. After the training, the farmer's land is surveyed. This survey is needed to investigate and gain certainty of the availability of such production factors as the following: size of land for cultivation; height of location levels and availability of family labour. The training courses are conducted in the raining season (from October to March) because farmers can successfully cultivate mulberry trees. The water is available, they can plant mulberry cuttings.

Extension services are an important element within the array of market and non-market entities and agents that provide human capital enhancing inputs, as well as flows of information that can improve farmers' and other rural peoples' welfare; an importance long recognised in development dialogue. The Goal of extension services in this project is to give information or solutions to farmers involved in sericulture. With this service the farmers are to be controlled and supported.

Credit system: The investment required for a sericulture production unit amounts to about USD 404 for small land, USD 600 for medium and USD 823 for larger land (table 2). Because the target group is defined as small farmers in the rural area the farmers' financial capacity is not given.

Table 2. Investments of sericulture production (in USD)

No	Description	Mulberry section	Silkworm section	Total
1	Small (0.30 ha)	120.86	283.33	404.19
2	Middle (0.50-0.90 ha)	201.44	398.51	599.95
3	Larger (1.00-3.00 ha)	400.00	422.94	822.94

The farmers would have to be provided with an adequate credit program. Credit supports farm development in at least three ways:

- Credit programs with low interest rates help small farmers to solve the capital problem.
- Credit programs help small farmers to get farm inputs to the farm.
- The farmers, who join a credit program, also get the training and consulting service.

Organisation of cocoon production: The following activities need to be organised:

- *Silkworm egg ordering*: The silkworm egg producer is a governmental breeding station and the eggs have to be ordered a minimum of two weeks before they hatch. The orders for eggs by the farmers ("plasma") are collected by the Non Government Organisation (NGO) staff and they ultimatively give these orders to the breeding station.

— *Early instar silkworms cultivation and distribution*: Egg hatching and early instar silkworm cultivation are very sensitive and require high hygiene. Therefore, the incubation chamber and all utensils have to be properly sterilised before beginning the process of incubation. The young silkworms should have uniformity of growth. Uniformity of growth is important for the next growth stages so that the mature silkworm can spin cocoons at the same time, and the farmers also can feed the mature silkworms and harvest the cocoons at the same time.

Distribution of young silkworms to the farmers is accomplished in the morning. It avoids high temperatures and overhead sun. Distances between farmers should be logistically well organised so that distribution runs efficiently. The hatch of eggs, young silkworm cultivation and their distribution require many operations and high costs, but it is important to do if the farmers are not sufficiently experienced. That system can be changed in future by the time the farmers have gained experience and are technically skilled.

— *Fresh cocoons purchase*: The fresh cocoons are sensitive to mechanical influences and can be broken easily. Therefore, cocoons should either be processed or dried within five days after harvesting. The farmers bring fresh cocoons in sacks and sell them to the NGO. Fresh cocoons will be 62 tested and selected. The price depends on the quality of delivered cocoons. Prices are fixed according to the following conditions: 1) the percentage of broken cocoons and 2) texture of cocoon skin.

— *Processing of cocoons*: The fresh cocoons can be processed to dried cocoons, raw silk, thrown silk and cloth. For efficient production the NGO organises the total cocoon harvest; in other words, the amount of cocoons that should be produced for every month.

Marketing

For sustainability of the sericulture project, silk products need to be sold. The silk products are dried cocoons, silk yarn and clothes. The products can be sold at domestic markets and export market. Export markets need large quantities and high qualities. Table 3 shows cocoon grading for international markets.

The price for an excellent cocoon quality is the highest cocoon price. The excellent cocoons have a filament length of minimum 1171 metres and a reelability of a minimum of 61 percentages. The lowest quality is the 4th grade. Standard quality is commonly defined as a grade of "2A".

To achieve the above named activities the NGO requires special personnel. The persons that should be employed are:

Project Manager: Projects are recommended to designate or hire a project manager with a background in sericulture production and project management to coordinate the team effort. The level of effort required is to be determined by local needs. Duties include coordination of all project activities thus, tracking and reviewing progress and ensuring that efforts are directed towards the project goals and coordinating project report preparation.

A project manager organises three sections: Sericulture, administration and processing of cocoons

Table 3. Cocoon grading

Results of cocoon filament length

Cocoon filament length (metres)	Less than 800	From 891 up to 960	From 961 up to 1030	From 1031 up to 1100	From 1101 up to 1170	From 1171 up to 1240	From 1241 up to 1310	From 1311 up to 1380	From 1381 up to 1450	More than 1451
Points	38.0	38.5	39.0	39.5	40.0	40.5	41.0	41.5	42.0	42.5

Results of reelability

Reelability (%)	Less than 39	From 40 up to 46	From 47 up to 53	From 54 up to 60	From 61 up to 65	From 66 up to 70	From 71 up to 76	From 77 up to 80	From 81 up to 85	More than 86
Points	47.0	47.5	48.0	48.5	49.0	49.5	50.0	50.5	51.0	51.5

Cocoon quality grading

Grades	4th grade	3rd grade	2ndgrade	1stgrade	Excellent
Points	Less than 87.5	88.5 88.0 89	90 89.5	90.5 91.0	More than 91.5

section. The sericulture section comprises responsibilities for training the farmers, young worm cultivation and distribution, extension services for the farmers. A responsibility concerning the administration section is for administration works, credits and purchasing of cocoons. The processing section includes responsibilities for cocoons processing, silk yarn and clothes.

Technical Support Manager: Future projects should employ core staffs that are informed in detail about the project activities of participating agencies. Assigning personnel to the project for a specified time period of two to three years during implementation helps to ensure they give the project proper attention, are accountable for progress and feel a sense of ownership in the project. Participating agencies must ensure that adequate staff time is allocated for each aspect of the project, including follow-up and report preparation, both key components of effective technology transfer.

The project's Technical Support Manager is responsible for mulberry section, young silkworm cultivation at "nucleus"-area and cocoon production at "plasma"-areas. He/she will be supported by field staff.

Trainer and field staff: Specialised agents commonly are employed after completing advanced formal training in their area of specialisation. To be effective, however, they also need competencies in the extension education process. In addition, they must understand the human development of learning, social interaction processes and they must become knowledgeable about the organisation within which they work.

Technical assistance: In order to produce high quality cocoons, the whole silkworm cultivation process must be done with care. One of the important factors is for farmers to cultivate the silkworms with using the methods given in the training course. The Banyumas Sutera Alam's supervisors should be providing assistance to farmers. Because the silkworm cultivation is a new activity, the availability of technical assistance is perceived as valuable because it can give farmers success in the project.

Administration and accounting personal: Scopes of work for these positions are to manage activities administration and the bookkeeping of all cash flows in the project.

Farmers

Target groups: The target groups are small farmers in rural area in Central Java, who have their own land for planting mulberries and have family labour available. The Project will impact directly on the disadvantaged farmers; -they are the villagers who are the underprivileged group of society eking out a subsistence level of living. They were born on the land. Their parents have no education, are illiterate, -their children often dropout of school at primary level due to lack of motivation.

The whole family can be involved in the cultivation of silkworm, grand parents, parents and children can all participate in the harvesting of the mulberry leaves and feeding the silkworms which is a task carried out 3 times per day and in the cleaning and disinfecting of the silkworm shed.

Promotion of Women: This project will also improve the life of women in the following ways:

— The cultivation of silkworm is not physically demanding work so village women who need to supplement the family income can carry it out.

— Women will be involved in the administration of the project especially in the running of the cooperation.

- In the cocoon sorting stage female labor is more suitable as it demands patience,—also in the spinning and weaving of the Silk, 90 % of the labor force is women.
- By offering working opportunities to the women it will increase their income and also create an incentive to avoid young marriages which is still a social problem in the villages.

Beneficiaries: As more farmers realise the potential in silkworm farming they will change from seasonal dependent crops to mulberry plantation. Within the first 5 years of the project 2500 farmers will be targeted with the training and support programs. Potentially 10.000 people can improve their conditions. Employment will also be available in the spinning and weaving and garment making when they come on line in 3-5 years from the start.

Activities of farmers

Mulberry Cultivation: The processes of planting mulberries are: preparing of land, planting of the mulberries cuttings, harvesting mulberry leaves and cutting of branches. The cuttings of mulberries must be free from diseases, be aged from three and half months olds and must come from prime seeds. Farmers directly plant the cuttings in the cultivated and fertilised land. The first harvest of mulberry leaves is up to six months after planting and leaves can be harvested every day for one month. The next harvest is three months after cutting of the branches.

Cultivating silkworm: Because of the high risks and the quite expensive tools and know how included, farmers start to rear silkworms in the 4th instar. Banyumas Sutera Alam does the silkworm rear until 3rd instar. Before the rearing of the silkworms, farmers empty and clean the rearing shed. Feeding the silkworms with fresh mulberry leaves is done three times a day: morning, noon and evening It takes around 20 days from 4th instar until the cocoons can be harvested. The silkworm rearing can be done for 8-10 periods per year.

Financial Possibilities

To commence a sericultural activity the farmers need a "kick start" capital. A small size farm needs USD 404, a middle size of farm needs about USD 600 and a larger size of farm needs USD 823.00. The need of financial support for sericulture investment can be reduced if the farmers themselves can supply parts of items in the investment, e.g. organic fertiliser, bamboos or woods for the racks and silkworm shed. Available family labour can be utilised to reduce about 20 percent of the budget. Mulberry cuttings could be supplied from the Banyumas Sutera Alam.

Banyumas Sutera Alam, which as a Non Government Organisation, has in this project the functions of coordinator, young worm producer and purchaser of fresh cocoons, needs financial support to realise the project. The construction part is including silkworm sheds for eggs-hatching room, for early instar silkworms and buildings for cocoon processing to silk clothes. The office equipment includes that necessary for silkworm rearing and for first cocoon processing equipments. Vehicles are required for transport of early instar silkworms to farmers, as well as motorbikes for farmers surveys. Training part is for costs of training 1000 farmers in house which means 3 days training and advanced training in fields.

The Funds for farmer are for 500 farmers in the form of revolving funds. Because the farmers' financial capacity is not sufficient, the farmers have to be provided with an adequate credit program. The credit programmes can be additionally provided from the following credit institutions: Government's Commercial Bank; Private Commercial Bank or NGO-International as funds.

Most of credit providers have difficulty in the process of administration, higher interest of loan and need collateral. This is the problem which prevents small farmers from getting credits. To avoid this problem Banyumas Sutera Alam looked to International Non Government Organisation—founder other for credits. Fortunately in 1998 to 2000 Banyumas Sutera Alam got support credits from Misereor, CRS, AusAID and DFID-British, both for the farmers and Banyumas Sutera Alam itself. For this funding Banyumas Sutera Alam has a duty to prepare the farmers for sericulture cultivations and provides revolving credits for the farmers.

Farmers who want to know more about silkworm farming have to register to join training held by Banyumas Sutera Alam. Banyumas Sutera Alam held three day training for farmers. Materials, travelling costs and accommodation were provided free. The training sessions are provided once a month in the rainy season (six months in a year). In the trainings and the information is given on the following:

- Mulberries cultivation: land preparation, mulberries planting, pruning, fertiliser, etc.
- Silkworm rearing: hygiene in the rearing house, feeding methods, harvest of cocoons.
- Building a silkworm shed/ rearing house.
- Cash flow calculation.

Farmers who are interested in the silkworm farming will be surveyed by Banyumas Sutera Alam officers to give possibilities for sericulture on their cultivated land. The farmers can apply for credits from Banyumas Sutera Alam if land and labor are available for the sericulture activities. The credit is given in a few steps. In the first month, farmers get credit for planting, cultivating the mulberries and organic fertiliser.

After about six months, farmers get the next credit for building materials for the rearing house and equipment for rearing. The amount of credit varies with the farmers. It depends on the farmer's application and the Banyumas Sutera Alam decision is based on the size of the farmer's land and his social economy. Credit repayment for basic instalment and interest rate 7.5%, paid in four years, including two years of grace period.

The credit repayment system is a direct repayment from their cocoons sale. The amount of credit repayment depends on the income of farmers from these cocoons sale, and agreement between farmer and Banyumas Sutera Alam. The farmers in one district should have a group and choose one of them to be a leading sericultural farmer. The group has functions to coordinate the activities and communication between Banyumas Sutera Alam and the farmers. The groups should have meetings periodically to discuss the problems and find solutions for themselves first.

Project Monitoring

Since the project has started in 1998 with experimental project until 2002 in the nucleus phase, the project was still on schedule. All these phases from experimental to industry phase have been completed. However, the project results were lower than expected.

Mulberry Section

Both quantity and quality of mulberry leaves are important as source of nutrition for silkworms. Quantity of mulberry leaves define how many boxes of silkworms and how many cycles could be reared in a year. As variables of quantity the mulberry leaves, have to be chosen as:

— Frequency of production cycle per year (periods p.a.), and

— Total of matured larvae of silkworm boxes per production cycle (box per cycle).

If the mulberry plants are well cultivated (pruned and fertilised regularly) therefore the mulberry plants are growing well, the frequency of production cycle per year could be high (10 times a year), and if the mulberry plants are growing well, leaves of mulberries are more available to feed silkworms, the total of matured larvae of silkworm boxes per production cycle could be also maximised (3 boxes/ hectare). The variety of mulberry plants is a variable of quality. There are five different varieties of mulberry in this project. However, the farmers cultivate those which are the most popular varieties: M. alba L(Ma), M. cathayana (Mc) and M. multicaulis (Mm).

The varieties of mulberry could present effects to quality of cocoons: total of weight cocoon, size of cocoon, percentage of shell etc. If the mulberry leaves has high quality of food source available to the larvae, the cocoons will be also in high-quality. These variables of quality and quantity influence the harvest and thus, the farmers' income. The income can be increased through improvement of quantity and quality of the mulberry plants (food source of silkworm larvae).

Silkworm Section

The quantity and quality of cocoons are influenced both from the mulberry plants and from the side of silkworm cultivation itself. There are a lot of variables that can be examined the out come, e.g. hygiene, condition of worm shed, disease control and feeding frequency etc. To examine these variables we chose only two important variables, namely:

— Fresh cocoon production per box (production of boxes in kg) and

— Price of fresh cocoons per kg (USD/kg).

At the mulberry section, frequency of production cycle per year was, we expected. These values were higher than criticality values of the project. On the other hand the total of advance silkworm boxes per production cycle had lower values than those that were designed. This result reflected that mulberry plants are growing still but they did not have many leaves for food.

In other words the mulberry plants were not well cultivated. This could have been for reasons that as they were not fertilised or not sufficiently pruned, etc. Because the mulberry plants were not growing well their leaves had poor quality and also were not sufficient to feed the worms. This condition evoked both lower values of fresh cocoons production per box and also lower values of cocoon quality.

The decrease of cocoon quality can be characterised with the price of fresh cocoons that had been lower than the expected price at USD 2.35. Apart from mulberry leaves the lower price could be caused by the other factors: e.g. feeding frequency, hygiene of worm shed, diseases, etc. To tackle above mentioned problems some additional management activities with additional costs of production are required, e.g. supplement of fertilisers, of hygiene and of management of diseases, repairing on rearing sheds to provide better environmental condition, etc.

Organisational Challenge

Besides variables of the mulberry- and silkworm sections, the other important factor that should be monitored is organisation of sericulture cultivate processes. For examples: education and training for the farmers; organisation of young worm transport, and management of farmers.

Training for the farmers

For the farmers sericulture was a new activity. Before, farmers did not have any knowledge of sericulture so at the first step they were trained in the Banyumas Sutera Alam project for three days. The training is continued if farmers are ready to cultivate silkworm (six month after planting of mulberry plants). This method is not sufficient for information transfer to farmers.

The farmers need even more time to practise and cultivate in their own rearing shed. To give more success BSA supervisors help farmers cultivation for first time and show practise methods to the farmers in their own fields. It was also difficult because the number of participants (farmers) rose in comparison to the availability of supervisors.

The distribution of young silkworm to the farmers should be done carefully and at the right time. The third instar silkworms which have to be distributed are in the phase of moulting to the 4th instar. At the time when they finish their moulting, they should have fresh mulberry leaves for food.

So the silkworms should be delivered at the right time and be fed immediately with fresh mulberry leaves. Some problems of distribution were that farmers were not ready to immediately pick up silkworms from the leader of their group or someone from Banyumas Sutera Alam who is delivering silkworms, because time of delivery was not fixed in exact hour unit.

For one reason, the limitation of transport for delivery and long distance from one farmer to the other. To solve these problems BSA tried to rear early instar of silkworms worms at the place of farmers. The leader or someone of the farmers will be trained how to rear early instar of silkworms. He rears them and distributes them to the single farmers. This method requires 107 time for training and new investments for the newly hatched silkworm sheds and the of equipment.

The farmers in one location are gathered into single groups. The leader of one group helps the new farmers and discusses about their experiences and problems with the sericulture activity. The meeting schedule is organised by BSA so that the BSA supervisor could be attending that appointed meeting. These methods certainly are effective due to giving information by supervisors from BSA and are consulted for problems from the group with single questions. The meeting should be held at least every month. If the meetings are regular, the problems of managing diseases of mulberry cultivation and silkworm rearing can be detected early.

Evaluation

In the Monitoring Section was shown that the project results were not as good as expected. In order to develop the project for the further future expected values of variables should be still achieved. To reach these values, following activities should be done:

— *Improvement of mulberry field*: regularity of fertilising and pruning, etc.

— *Improvement of silkworm rearing*: minimum feeding frequency is three times a day, using more Morus alba variety, better hygiene in rearing sheds, etc.

— *Improvement of organization*: early instar silkworm distribution, regularity in visiting farmers by supervisors, etc.

The project still has IRR higher than reference interest of the bank. The middle land has the best results than the other two sizes land, because middle land had better results in the last years project in comparison with the small and larger land. To increase values of IRR and NPW, farmers should develop their income. The income can be increased with developing higher yields of fresh cocoons per box, both quantity and quality of cocoons.

The income can not be increased within periods of silkworm rearing. The expected value of period of rearing of silkworms per year was at maximum value: 10 time periods per year. So the expected value of fresh cocoons yield per box is 25 kg, as one possibility which can be developed. Theoretically, fresh cocoons yields can be increased to above 25 kg per box. One box eggs contains 25,000 eggs. They hatch and develop to become mature silkworms ready to spin cocoons. The weight of one fresh cocoon is about 1.5 to 2.0g.

If mortality rate of worms is 25 percent so still 30 kg of fresh cocoons per box is feasible. The other possibility to increase income is through quality. The market value of fresh cocoons is depending on percentage of broken cocoons and contents of silk. The sell price was expected with Rp.20,000 (equals USD 2.38) in condition 6 percent of broken cocoons and 23 percent contents of silk.

It is possible to reach a higher market value if farmers rear silkworms with more intensively, e.g. more feeding, better hygiene in rearing sheds etc. But it is definitely more difficult to get a percentage of silk of more than 23 percent because contents of silk also depend on the breeding of silkworms. So farmers have more possibilities to increase the market value of cocoons through decreasing percentage of broken cocoons. Broken cocoons can be reduced if the silkworms, at the cocoon processing phase, are handled carefully, e.g. the only collection of silkworm that are ready for cocooning; using optimum seriframes for processing of cocoons.

To increase income through setting up the aspect of quality is rather difficult. But it could be more feasible for the next future that farmers reach better experience of sericulture rearing in future. To develop the ongoing sericulture project at BSA there should also be improvement done through rearing assistance of sericulture at the farmers' side, and BSA as nucleus of the project, should be enhanced as well.

SWOT-Analysis

SWOT-Analysis is a systematic identification and a strategic decision making operation. The aim of SWOT is to develop a plan that takes into consideration many different internal and external factors and maximises the potential of the strengths and opportunities while seeking minimise the impact of the weaknesses and threats. The internal variables are composed of variables such as: management, structural organisation, production and operation, finance and human resources. Internal data is found from inside the company itself, e.g.:

— Financial report (balancing, cash-flow, cost benefit, and structural fee)

— Human Resources reports (total employees, education, skill, experience, salary, and turnover).

— Operational Activity report

— Marketing Activity report.

External variables are outside of company control and the company has no power to control these external variables such as:

Macro Variables. A macro variable is a large social power and can influence all in the micro industry such as: economic, politic and government, culture, demography, environment and technology.

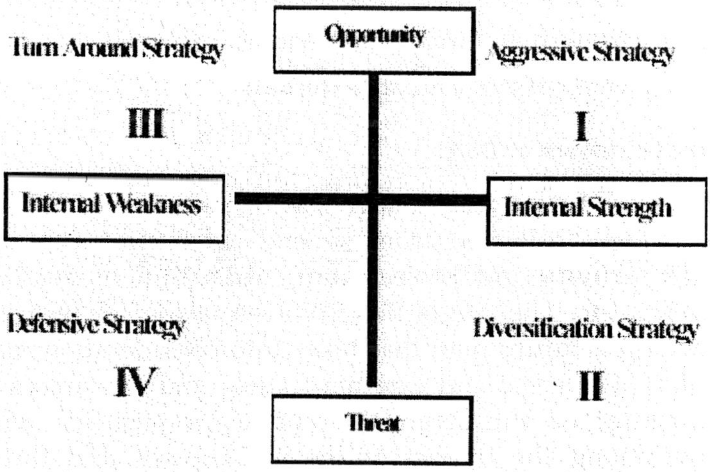

Figure 2. SWOT Analysis

Micro Variables. Micro variables are those involved in the environment that have a first hand job in influencing the market industry, through costumer bargaining power, supplier bargaining power, new comers threat, competition between similar companies, and substantial product. However, external data can be collected from outside company resources such as: market analysis, competitor analysis, community analysis, supplier analysis, government analysis, and other important community group analysis.

The logical base idea from SWOT is to maximise strength and opportunity, beside it seeks to minimise weakness and threats. In other words, SWOT-Analysis a situation analysis because Business Practitioners must analyses factors in company strategy (Strength, Weakness, Opportunity, and Threat). SWOT-Analysis strategy considers four conditions:

1. Strength & Opportunity (internal)
2. Opportunity & Weakness (external vs. internal)
3. Strength & Threat (internal vs. external)
4. Weakness & Threat (external)

SWOT-Analysis of BSA

The SWOT-Analysis of BSA was conducted in 2003. The data were collected from several activities: interviews with farmers and project staff, utilisation of available reports, etc. The aim of this SWOT-Analysis for BSA is to develop a future plan that takes into consideration many different internal and external factors, and maximises the potential of the strengths and opportunities while minimising the impact of the weaknesses and threats from BSA itself. With this analysis and improvement of sericulture in the farmers, the project sericulture at BSA can be developed for the future.

Strengths

Get the support from society: Although BSA is a relatively young NGO, nevertheless, the program of sericulture has involved a lot of farmers and women workers at cocoon processing. With this support it is still possible for BSA to develop furtheron sericultural area in the future.

Family labour available: The most of farmers has a family with 2 to 4 children. Therefore, family labour at the farmer level is available.

Existence of cooperation with national and foreign institution: BSA has good cooperatives with Universities, Research Institutions and local Government in Central Java. Furthermore BSA also has good cooperatives with international NGOs: AusAID, CRS, DFID-British Embassy, Misereor-Germany, etc. This cooperation provides chances for BSA to get support in technologies (through training, seminars, etc) and financing for developing of the project.

Facility of sericulture processing: BSA has the complete machinery for silk processing. These machines have high capacity. BSA can buy cocoons from the farmers with high capacity.

Hand made of silk products: The silk clothes are made with traditional machine (ATBM = Alat Tenun Bukan Mesin). These clothes are exclusive and of high value.

Weakness

Fluctuation of silkworm quality: The silkworms have fluctuation of quality. These are caused by quality of eggs, newly hatched silkworm rearing, distribution of silkworms—so that the production of cocoon can be influenced. The cocoons production are lower.

Irrigation not possible at the farmers: Most of the farmers do not have irrigation on their land. They can not rear silkworm in dry season, especially in time from August to September, because mulberry leaves are not available.

Limited of fund: Ability of investment fund is very limited so that require long time to extend business.

Limited of marketing experts: Availability of expertise in marketing are still few. Increasing of the product's volume could present difficulties at product marketing.

Opportunities

Interest of farmers to sericulture still high: It is still possible to expand the project to the farmers. Most of farmers want to join the project. It can be seen if Banyumas Sutera Alam offers the training, the number of participants will be high. Improving of silkworm eggs through government research or university Government research institutions and Universities are doing research in sericulture every year. Banyumas Sutera Alam can provide information; know how; eggs from these researches. Existence of training offer from local government and all donor BSA has training offers from all donors and government each year. This could give chances for BSA to improve its capability.

Market of fresh cocoons and silk products: The market is still open with high capacity, especially for cocoons. So that BSA has still chance to market cocoons.

Threats

Farmer knowledge is low: Because target group are small farmer in rural region the farmers are not well educated. BSA should give information through more practice and activity. The transfer of know how needs more time and patience than in cases of educated people.

Monopoly of silk eggs production: The government has the monopoly of the egg production. Hence, it is difficult to get eggs from another source.

Cheaper cocoons from another country: The world market price of cocoons is lower than Banyumas Sutera Alam, especially those from China. Therefore Banyumas Sutera Alam can market their cocoons if the price is lower or equal with the current price.

The lower price of synthetic silk like fabrics: The price of synthetic fabrics that are silk-like is cheaper than pure silk. It is difficult to compete with the synthetically produced fabrics.

The forming of society opinion about story; level of silk softness (synthetic silk): Most people know that silk means brightness, very soft texture and cheap product. The image of synthetic silk is stronger than pure silk. So that people want to buy only synthetic silk which fulfils their expectations. Based on the aspects resulting from the SWOT-Analysis a TOWS-matrix was developed.

A combination of several external factors (threats and opportunities) and internal factors (weaknesses and strengths) lead to the composition of certain strategy-elements. These are classified in short- (combination of strengths and opportunities), middle- (combination of strengths and threats as well as weaknesses and chances), and long-term (combination of threats and weaknesses) elements. These elements of strategy were combined to three complete strategies:

1. The strategy for breeding of eggs;
2. The strategy for sericulture cultivation at the farmers;
3. The strategy for marketing.

The strategy for breeding of eggs

The problem of the eggs' quality can be handled with following strategies:

SO- Strategy:	Invent innovations by applying silkworm techniques.
WO- Strategy:	Seek and apply new egg strains with a better quality through government and Universities research
ST- Strategy:	Establishment of breeding research with institutions and founders.
WT- Strategy:	Providing with other sources of silkworm eggs

The strategy for sericulture cultivation at the farmers

SO-Strategy:	Improvement of better sericulture activities with farming sample
WO-Strategy:	Utilising the training program to produce expertise.
	Increase the farmers' awareness for the importance of silkworm farm operations.
ST-Strategy:	Skills improvement of farmers through a training program which is held by BSA.
WT-Strategy:	Forming of groups of silk farmers to assist the increase the farmers' ability.
	Study comparison and cooperation with other institutions.
	Propose of long-term loan for the expansion of the effort.

The strategy for marketing.

SO-Strategy:	Conservancy quality product and quality service to form the positive image for company.
	Make silk material to become silk textile.
	Cooperation with government institution or school for making of batik uniforms.
	Expand market compartment to improve marketing.
WO-Strategy:	Optimising the production volume by increasing employees' skills.
	Improve marketing promotion effectively and efficiently.
ST-Strategy:	Maintaining quality product to face the competitor.

Increasing the cocoon production volume.

Determination of market goals (sale position outside area of product sale of non-pure silk).

Socialisation of pure silk product

Existence made in non-machine-weaving-tools (ATBM).

WT-Strategy: Recruit expertise in marketing area to increase the company's performance

13

Scenario of Indian Silk Industry

Bonded Child Labour

Children are working at every stage of the silk industry. These children work twelve or more hours a day, six and a half or seven days a week, under conditions of physical and verbal abuse. They suffer injuries from fumes, machinery, sharp threads, boiling water, and dead worms. Girls face particular abuses, including sexual abuse by employers.

Thread-Reeling and Twisting

Mature cocoons are transformed into silk thread through a process of reeling and twisting. Children are cooking cocoons, picking out the dead worms, twisting the filaments into thread, and helping with odd jobs. Reeling is the process in which dried cocoons are turned into hanks of multi-strand yarn. The first stage of the reeling process is called "cooking"-boiling the cocoons. The cocoon consists of a single long cluster of silk fibers held together by the water-soluble protein sericin. Boiling the cocoon dissolves the protein.

Children stand over open basins of boiling water that are heated below by charcoal or fire, keeping the cocoons immersed in the water with a ladle and palpating the boiling cocoons with their hands to see if the sericin is dissolved and the fibers loosened enough to be unwound. The children are required to reach into the boiling water to feel the cocoons with their hands, rather than with a spoon, on the theory that their hands can more accurately tell when the cocoons are ready to reel.

After reeling, individual threads are twisted into strong, multiply silk thread. Twisting usually takes place in small factories with between fifty and a few hundred spindles. These factories use bonded child labour. Children stand for long hours loading the spindles, guiding the thread, and performing other tasks.

Children typically reported working ten to fourteen hours a day, with breaks to eat, six days a week. Children will work for even longer when the demand for thread increases; for example, before a festival when women purchase new saris, factories will sometimes operate twenty-four hours a day. The supervisor at a silk twisting factory told Human Rights Watch, "Outside employers are taking business away. We have to work very fast-day and night. " Thirteen-year-old D.G. Sagar, who was Dalit, said that when he worked at night, "I didn't like working in the morning, but the owner would insist. He demanded that I come in the morning because I was a good worker. Because the owner was so insistent, I would take the extra work and work day and night."

Most children working in reeling units are fired at about age fifteen or sixteen and replaced with their younger brothers or sisters; some will go to work in twisting units, but most must seek other work. Thus whatever skills they may have acquired in the silk unit do not help them in adulthood. Indeed, the injuries and illness suffered as child workers limits what work many can do as adults.

Working Conditions

From immersion in scalding water and handling dead worms, reelers' hands become raw, blistered, and sometimes infected. Twelve-year-old T. Basheer worked in a silk reeling unit until twenty-five days. At first he picked worms, then he graduated to cooking. "Boiling water falls on your hand," he told us. "You are always in water, standing in it. The skin on your hands and feet peels off. It gets loose." Anesha K., eleven years old, started picking worms when she was nine and now she had been at an NGO-run residential school for four months. She showed lumpy scars on her hands and explained: "I didn't like working because my hands would get infected. I got holes in my hands because I put them in the hot water and then they got infected. I couldn't eat. I had to eat with a spoon." Anesha K.'s shins, ankles, and feet were covered with burn scars from boiling water. Sericin vapours from the boiling cocoons, smoke, diesel fumes from the machines, and poor ventilation cause respiratory ailments such as chronic bronchitis and asthma. Children also suffer cuts to their hands from the threads and from the machines while winding and twisting the silk thread.

P. Kattaraman told Human Rights Watch: "The machinery was bad-it can cut your fingers. If you cut your finger, even then, you wouldn't get sent home. They would put coffee powder on it and you would keep working. There wasn't any medicine." P. Kattaraman showed us the bends of his fingers where the threads would cut. Thirteen-year-old D.G. Sagar who worked at a silk twisting unit from age five to age eleven said: "When I was working at night, the skin would come off of my fingers. I would get very sleepy and miss, and the machine would take off the skin." Standing at the twisting machines all day without rest or carrying ladles of hot cocoons, basins of dead worms, and other heavy materials causes children back and leg pain, and some suffer leg deformities, including bowleggedness.

Children also complained of cuts and other injuries from threads and machines; headaches; pain in the back, legs, neck, and abdomen; colds and bronchitis; hearing loss; and vision problems. The noise from the twisting factories Human Rights Watch visited was deafening, and employers often play loud music in the mistaken belief that it will prevent hearing loss from the machines. Dim lighting inside the factories and the close eye children must keep on the threads and machines also causes vision loss. When children are injured or ill, they often must continue to work or lose the income, which forces them deeper into debt.

Employer Abuse

In addition to injuries related to the work itself, children reported being beaten and verbally abused by their employers and older workers. Nine-year-old P. Ningamadiah, said: "At work the supervisor used to beat me with a belt. He tied me up and beat me with a belt on my back. He did this two or three times. . . . He tied a chain that was attached to the wall to my leg." His owner beat him, he said, "If I didn't do my work properly." T. Basheer, age twelve, told us: "If we make some mistake, the owner beats us on the neck with a stick. If we get sick, the owner comes to our house, uses bad language, and takes us to work. He comes and scolds us and takes us back because he doesn't have a replacement." S. R. Kollur, a twelve-year-old boy who worked from age ten to eleven in a silk twisting unit, said: "Once, my legs were hurting, and I wanted to sit down. I cannot stand for a long time because it makes my legs hurt. I went to rest, and the owner kicked me in the legs."

An estimated 200,000 children under age fourteen are working in the silk industry in the Varanasi area, about half of whom are bonded to a nonfamily member. As already noted, approximately 40,000 to 50,000 children work in bondage on silk handlooms in Kanchipuram. Until around age fourteen or fifteen, children are usually too small to operate the looms themselves. Instead, they assist weavers by sitting next to them at the loom, legs dangling in the pit, catching the shuttle and embroidering the border.

Girls and women also wind the weft thread onto bobbins that are loaded onto the shuttle and woven through the warp threads. In Varanasi town, in nearby Dalit villages, and in Kanchipuram, girls and boys working at the looms, assisting weavers and winding thread on bobbins.

In both Varanasi and Kanchipuram, even if they attended a night school, children reported working twelve to thirteen hours a day with breaks for breakfast and lunch. In Kanchipuram, children reported working seven days a week with two half-days off a month.

Girls rarely become weavers, which limits how much they can earn. Researchers in eastern Uttar Pradesh, where most silk weavers are Muslim, have found that girls' labour is nearly invisible. In this area, girls work inside homes filling bobbins and embroidering saris; when they work for someone outside the family, the work is still brought home. In contrast, in Kanchipuram, Human Rights Watch interviewed girls who worked both in others' homes and in factories. However, girls there are still much less likely than boys to become weavers or loom owners.

When Human Rights Watch asked a group of weavers, assistants, and loom owners in Kanchipuram about women's work, one man responded: "Women wire the silk-they put it on bobbins and roll it. They don't weave as weavers. They do assistant work." According to another man, "10 percent of women are weaving. The rest do assistant work. Women can get Rs. 300 [U.S.$6.25] like the boys. Most are working at home. If they are weaving, they get the same as a man." Although adults can, and sometimes do, work as assistants, weavers prefer children. National Labour Institute researcher Babu P. Remesh, in a survey of labour in the Kanchipuram silk industry in 2000, found that weavers prefer children for the following reasons:

(a) the recurring labour costs in the case of adult helpers are high (for an adult helper, the monthly wages are about Rs. 600 [U.S.$13], where as for the child labour it is about Rs. 300 [$6.25] or even less than that);

(b) the perception of the weavers that the child helper remains for longer periods with them compared to the adult ones;

(c) the belief that children are better than adults (nimble fingers argument)-which in reality implies that they could be better monitored and exploited.

Unlike children bonded in silk thread production and certain other industries such as carpets, most weavers' child assistants continue to work in the industry as adults. Therefore, parents may also bond their children to a weaver in part for training, believing that children must learn weaving early under a "hard task master." According to Professor B.N. Juyal, an expert on child labour in the silk industry, while poverty is one reason children enter the weaving industry, other major causes are the great demand for children's labour, the belief that children are learning valuable skills, and socialisation in the "traditional craft culture." Children's Testimonies

In a Dalit and low-caste neighbourhood less than two kilometres from the deputy labour commissioner's office in Varanasi, Uttar Pradesh. Rustamlal M., who was twelve, said he attends school for four hours in the morning, then works six and a half or more hours at the loom.

Fourteen-year-old Ashish M. told us that he could not leave his loom owner because he was paying off an advance, which in two years he had reduced from Rs. 2,500 (U.S.$52) to Rs. 475 ($9.90). "The owner pays but deducts for the advance," he said. "He deducts but won't write off the whole advance. . . . We only make enough to eat."

Since then, every morning, seven days a week, he has woken up at 7:00 or 7:30 and gone to work making borders for saris. He told us that he works for twelve hours, with an hour break to go home for lunch. He makes Rs. 300 (U.S.$6.25) a month, but, he explained, some of that is taken out to pay the loan. He said he still owes Rs. 400 (U.S.$8.33) but that the family has borrowed more since the original loan. His boss's children go to school, he said, but he does not.

Nesan S., also ten years old, said he had been working in Kanchipuram, Tamil Nadu, assisting a weaver for the last two years. He told us that he earns Rs. 200 (U.S.$4.17) a month and is paying off a Rs. 6,000 ($125) advance that his family used to buy a loom on which his parents now work. "My father asked me to stop studying. I went to third standard [third grade], then I stopped," he explained. Nesan S. said he goes to work at 7:00 or 7:30 in the morning and has two half days off a month.

Bondage in Silk Weaving

Like many industries, the silk handloom business is structured vertically. Children are at the bottom, initially bonded for around Rs. 2,000 to 10,000 (U.S.$42 to $208). Children work for weavers who are themselves typically in debt to a middleman (sometimes called a "master weaver") or directly to a trader (sometimes called a "gaddidar") and who are not free to work for any other. The trader, who actually sells the sari, takes most of the profit; in Kanchipuram a sari sells for almost double the total production cost.

A Varanasi silk trader who said that he usually works directly with weavers said, "We sometimes go to middlemen if there is high demand. The middleman takes half the profit from the weaver. Our costs are the same but he takes it out of the people beneath him. The middleman gives the advance." In addition to holding a debt, traders and middlemen can control the weaver by owning all or key parts of the loom; supplying the thread, zari (gold thread), and design; and marketing the sari. Human Rights Watch's findings of pervasive bondage in silk weaving are supported by extensive survey research conducted by the National Institute of Labour.

A medical examination at a health camp for some eighty-five children released from bondage in Kanchipuram's silk looms in November 1997 found that all were malnourished, two had pulmonary tuberculosis, twenty-six had various skin infections, and many had vision, dental and hearing problems. Researchers in Uttar Pradesh have found similar problems among children working on silk looms in that state. For example, a worker with a community health organisation in Varanasi told Human Rights Watch that silk workers report breathing problems and pain in their shoulders and upper backs.

Employer Abuse

Many children reported that loom owners and weavers abused them verbally and physically. According to a fourteen-year-old bonded boy who worked as a weaver's assistant in Varanasi: "If we don't do the work, they will beat us. They curse us and beat us." Twelve-year-old Kanshi Ram, who works as a weaver's assistant, told us that if he gets sick, "the owner will scold me and will not allow me to leave. If I take leave he will beat me. He will ask me if I am cheating him."Nallanayaki P., who was thirteen years old and from a low caste told us: "Always [the weavers and owners] are beating me-I don't like to work. They always scold and shout. They beat me on the back and head. They are always knocking their fists on my head or hitting me with a comb [wood piece in the loom]."

Caste-based Discrimination

Allocation of labour on the basis of caste is one of the caste system's fundamental tenets. Within the caste system, Dalits, or so-called untouchables, have been assigned tasks and occupations deemed ritually polluting for other caste communities. Most bonded labourers are low-caste, illiterate, and extremely poor, while the creditors/employers are usually higher-caste, literate, comparatively wealthy, and relatively more powerful members of the community. According to government figures, 86.6 percent of bonded labourers are Scheduled Castes and Scheduled Tribes. In sericulture, most bonded children are Dalit or Muslim, depending on the area.

The traditional silk weaving caste is a lower caste, called a "backward caste," but in both Varanasi and Kanchipuram districts, Dalits have entered the weaving profession in significant numbers, often by being bonded as children. Muslims also dominate sari weaving in some areas, including in urban Varanasi. However, in Varanasi district, Dalits and lower castes have begun weaving in increasing numbers, in part because communal violence and attacks on Muslims have pushed them out of the profession. Dalits and low-caste Hindus are the most vulnerable to bondage for the following reasons:

— upper castes traditionally expect that Dalits will perform free services, which helps to sanction the bonded labour system and results in poor Dalits being bonded when poor caste Hindus are not;

— Dalits are typically landless and therefore economically dependent on their employers. Economic dependency also keeps them from reporting atrocities against them;

— upper castes levy social and economic boycotts as well as perpetrate violence against Dalit communities who challenge traditional practices and try to assert their rights;

— upper castes dominate local political bodies, the police and the judiciary, bonded labour vigilance committees, and child labour committees that are supposed to enforce the law on bonded and child labour; and

— discrimination against Dalit children in school encourages them to drop out and be sent to work.

Dalits and lower castes are typically restricted to tasks and occupations that are deemed too "filthy" or "polluting" for higher-caste communities, and the poor remuneration of manual scavenging, agricultural labour, and other forms of low-caste employment often forces families of lower castes into bondage. For example, Human Rights Watch interviewed Dalit villagers who weave saris in their homes on looms owned by traders, but whom local landlords also force to work their land. "We have very little land, less than five acres," a Dalit woman in the village told us. "Yes, of course we work on the landlords' land." In exchange for a day's labour, a worker receives five kilograms of wheat, worth about Rs. 40 (U.S.83¢). "They don't even measure the five kilograms," one man complained. "They just fill up a sack and bring it out to us." Another man explained that they couldn't survive on the money earned from this and from sari weaving, so they had to take loans from the traders. Because upper castes have expected and extracted free services from Dalits, Dalits are more likely to become bonded compared with equally poor higher caste Hindus.

Most Dalit victims of abuse in India are landless agricultural labourers who form the backbone of the nation's agrarian economy. Despite decades of land reform legislation, over 86 percent of Dalit households today are landless or near landless. Those who do own land often own very little. Land is the prime asset in rural areas that determines an individual's standard of living and social status. Lack of access to land makes Dalits economically vulnerable; their dependency is exploited by upper- and middle-caste landlords and allows for many abuses to go unpunished.

Landless agricultural labourers throughout the country work for a few kilograms of rice or Rs. 15 to Rs. 35 (U.S.32¢ to 75¢) a day, well below the minimum wage. Many labourers owe debts to their employers or other moneylenders. Thus, although poverty plays a significant role, connection between caste and bondage goes far beyond poverty to what Professor Thorat has described as the "extra economic compulsion of caste." Extreme economic and political measures are meted out for transgression of caste norms. Bonded labour flourishes in this environment because people cannot complain.

Caste-based violence is directly linked to child bondage. Researchers with the Tamil Nadu-based NGO Legal Resources for Social Action (LRSA) report an increase in child labour, especially in domestic and hotel work, following upper-caste raids on Dalit villages. When the families' economic assets are destroyed, the parents are more likely to need the children to work. In addition, [i]nvariably, their school or any common asset-e.g., village administrative office, post office, T.V.-they are all put in the main [upper-caste] area of the village so when there is a clash, . . . kids can't travel safely to the main village. . . . There are many occurrences where Dalit children stay out of school for one to three months because, for example, of a riot in the village or reprisals against Dalits, and police take men away . . . . Children are stranded . . . There is a decreased interest in going to school, and they need the income.

Researchers at the National Labour Institute have confirmed these findings. Discrimination against Dalit children in school also encourages them to drop out and go to work. Most of the government schools in which Dalit students are enrolled, where they exist at all, are deficient in basic infrastructure, classrooms, teachers, and teaching aids. Dalit students sit at the back of the class and are often treated badly by upper-caste teachers and staff. A majority of Dalit students are also enrolled in vernacular schools whose students suffer serious disadvantages in the job market as compared to those who learn in English-speaking schools.

Usually a non-Dalit person is providing the food because non-Dalit parents won't want their children to take food from Dalits. This person calls them by their caste name and makes them stand in separate queues. In Uttar Pradesh, documented discriminatory practices against Dalit children in schools include:

— discrimination against Dalit settlements in the location of schools;

— teachers avoiding physical contact with Dalit children; children from particular castes being special targets of verbal abuse and physical punishment by the teachers;

— low-caste children frequently being beaten by higher-caste classmates.

Discrimination combined with low returns on education, including discrimination in employment, encourages Dalit children to drop out of school; Dalit children drop out of school at a much higher rate than non-Dalits, and there is a higher rate of illiteracy among Dalits than among non-Dalits. According to Gilbert Rodrigo, "Dalit kids get left behind, and when they don't see graduates getting jobs, they loose interest." A group of low-caste and Dalit silk weavers told Human Rights Watch that if people went to school, "10 percent go for jobs after they study. The rest go back to the first job. Even that education doesn't help." And a low-caste teacher in one of Kanchipuram's night schools described his own school experience: "My class had thirty students. I was the only one who went to another job. The twenty-nine others are weaving." Although some allege that Dalit and low-caste parents do not value education, this is not surprising if the quality of education offered is very poor and if they see no returns on it.

Role of Government

India's national and state governments share responsibility for protecting bonded child labourers. Their most basic obligations are to prevent children from becoming bonded in the first place, to remove those who are from bondage, and to prevent them from becoming bonded again. On paper, the government has developed a wide array of laws and policies on bonded labour and child labour, which in theory should protect bonded children. In practice, these are never implemented for most bonded children. Without effective enforcement of either bonded labour or child labour laws, children in bondage-the most vulnerable of all working children-are without recourse.

In the mid- and late 1990s the Indian government, under scrutiny from local NGOs and international organisations, paid increased attention to child labour and, to a lesser extent, bonded labour. In August 1994, then-Prime Minister P.V. Narasimha Rao announced an initiative to bring two million children out of hazardous labour by 2000. In December 1996, the Supreme Court outlined a detailed framework for punishing employers of children in occupations deemed hazardous by law and for rehabilitating the children.

In 1997 the Court ordered the National Human Rights Commission (NHRC) to supervise states' implementation of the bonded labour law, and the NHRC then began appointing special rapporteurs who applied pressure in certain regions and industries. There were some high-profile raids on employers, and a few were prosecuted. Through the efforts of both government bodies and NGOs, public awareness that some forms of child labour and bonded labour are illegal increased. While states continued to deny the presence of bonded labour in their territories and grossly underestimated the use of child labour and bonded labour, the steps taken were promising.

However, by 2003 most government promises have not materialised. States are still rarely freeing and rehabilitating bonded labourers, and the central government, with the exception of the NHRC, is acquiescing. Almost all government officials whom Human Rights Watch interviewed for this report denied that children were bonded. Both the central and state governments are backing away from their limited efforts to enforce the child labour law, and, with the notable exception of a few individuals, most government officials with whom we met took a dim view of it. Some argued that families need the children's income and that children would be pushed into more marginal and dangerous occupations. Others contended that law enforcement was irreparably ineffective and, therefore, not worth pursuing, even when they were themselves responsible for ensuring effective law enforcement.

Rehabilitation programmes for both bonded and child labour remain promising but extremely limited. Tellingly, eight years after former Prime Minister Narasimha Rao's announcement, the secretary of the Ministry of Labour told Human Rights Watch that the goal of removing two million children from hazardous labour by 2000 had been pushed back to 2005.

Child labour and bonded labour are treated entirely separately in law and in government policy, although both regimes are based on punishing the employer and rehabilitating the worker. The child labour regime revolves around the distinction, enshrined in the Child Labour (Prohibition & Regulation) Act, 1986, that some work is hazardous to children and some is not. For children under fourteen, the law prohibits work deemed hazardous and regulates everything else; children employed illegally are to be withdrawn from work and, following a 1996 Supreme Court order, rehabilitated. Employers who violate the law are to face fines and imprisonment.

All aspects of the silk industry are considered hazardous to children under age fourteen. Children age fourteen and older are not protected by the child labour law. The bonded labour regime is based on the same tenets as those of hazardous child labour: finding, freeing, and rehabilitating the bonded worker, and punishing the employer. The Bonded Labour System (Abolition) Act, 1976, which prohibits all forms of bonded labour for people of any age, establishes this framework.

However, compared with the child labour law, the bonded labour law allows for a longer maximum prison sentence, protects children over as well as under fourteen years old, extinguishes the debt, and actually codifies the rehabilitation requirement. In addition, offences under the bonded labour law, unlike the child labour law, are cognizable, meaning that a police officer may arrest without a warrant and initiate prosecution. Despite these two sets of laws, there are no specific programmes for bonded children, who fall at their intersection.

Bonded Labour

In contrast with extensive rhetoric on child labour, the government continues to pay little attention to bonded labour. Although there are an estimated twenty to sixty-five million bonded labourers, adults and children, in India, the government identified only 1,795 bonded labourers from 2000 to 2002, according to its own records, and fined or jailed almost no employers. High-level government officials adamantly deny that children are bonded at all, but fail to collect basic information, including age, on bonded labourers. An exception is the National Human Rights Commission (NHRC), which the Supreme Court ordered in 1997 to supervise the 1976 Bonded Labour System (Abolition) Act's implementation. The NHRC has been able to pressure some local officials into freeing and rehabilitating bonded children in some areas, primarily in Uttar Pradesh's carpet belt. But its resources and power are limited.

States are responsible for enforcing the Bonded Labour System (Abolition) Act, 1976, and do so through their district magistrates, in some states called district collectors or deputy commissioners, who are directed to form bonded labour "vigilance committees." The central government is responsible for ensuring that states enforce the act and that the vigilance committees are formed. The central government funds state surveys of bonded labour, evaluations of the bonded labour law's implementation, and public awareness campaigns. It also provides half of the funds for rehabilitation assistance, which is currently set at a total of Rs. 20,000 (U.S.$417) per bonded labourer, slightly less than the annual per capita income. District magistrates are appointed civil servants and are the top authorities at the district level.

They oversee government administration, including the administration of justice, and some fifty to sixty distinct departments. Their wide array of duties includes identifying cases of bonded labour in their districts, freeing the labourers, initiating prosecutions, making sure available credit sources are in place so that freed labourers will not be forced into bondage again, and constituting and participating in the vigilance committees. The vigilance committees are charged with advising the district magistrate to ensure that the bonded labour law is properly implemented; providing for the economic and social rehabilitation of freed bonded labourers; coordinating the functions of rural banks and cooperative societies to help ensure freed bonded labourers have access to credit; monitoring "the number of offences of which cognizance ought to be taken under the act"; and defending freed bonded labourers against attempts to recover the bonded debt.

In 1996 Human Rights Watch found that very few vigilance committees had even been formed and was unable to find any district in which such a committee was operative. In April 2002, officials in the Ministry of Labour told Human Rights Watch that vigilance committees had been created for every district. But regardless of whether they exist in name, almost none are actually functioning. According to NHRC Special Rapporteur Chaman Lal, the commission investigated and found that the vigilance committees "were non-existent or defunct and, where they had been revived, they were not doing work." The commission has been trying to revive them, he said. According to a highranking government official in Tamil Nadu:

On paper, the vigilance committees look good. But they are packed with people who belong to the ruling party. It is a status symbol to be a member. Every district collector has learned that to form a committee, he has to consult with the [party] Secretariat. Even S.C./S.T.s [Scheduled Caste and Scheduled Tribe members] will still belong to the ruling party so they toe the line. . . . Until now, the committee hasn't been involved in rehabilitation. The stereotype is that there will be a ten minute long meeting just to say that there is no bonded labour.

Government Failure to Protect Children with the Bonded Labour System (Abolition) Act, 1976. High-level officials responsible for enforcing the bonded labour law categorically denied to Human Rights Watch that children were bonded, especially in the silk industry. Many government officials interviewed by Human Rights Watch considered the bonds, which are legally extinguished by the Bonded Labour System (Abolition) Act, 1976, as enforceable debts. Worse, police in some cases are reported to have forcibly returned bonded children to their masters. Denial at all levels of government that children are bonded and the consequent failure to apply the bonded labour law for children creates a fundamental barrier to freeing and rehabilitating children in bondage.

Some officials in interviews with Human Rights Watch did not seem to understand what constitutes bonded labour. The Varanasi deputy labour commissioner told Human Rights Watch: "In the silk industry, it's all neighbourhood children so they can't make them bonded." The district collector of Kanchipuram, Tamil Nadu, said of children working on silk handlooms:

> Usually a parent takes several thousand rupees in loans and lets the child work to repay them-it's a kind of bonded labour. Just to repay the debt is technically bonded labour. . . . If they borrow money, it has to be repaid. There is a quid pro quo. So unless there is a physical instrument, it can't be classified as bonded labour. Technically, underpayment is bonded labour, but they don't complain and unless they complain, it can't be identified.

The then-labour commissioner of Karnataka, E. Venkataiah, when confronted with around eighty bonded children in Magadi's silk twisting factories, said to the press in January 2002 that "we can't do anything about the loans they have availed of." These are clear misstatements of the Bonded Labour System (Abolition) Act, 1976, which does not require any physical instrument for bonded labour to exist and which extinguishes bonded labourers' debts. A Tamil Nadu government official, who wished to remain anonymous, confirmed that many district magistrates (district collectors) do not understand, or disagree with, the definition of bonded labour and, therefore, do not enforce the law:

Bureaucrats are used to seeing in black and white. If it's gray, unless you are motivated, you won't take corrective action. You will take the easy way out. For example, the definition of bonded labour. They are bureaucrats. They see it as parents asking for money, so why penalise the employer? They are not able to understand. They see it as the employers give money when people need it. They only see bonded labour as being tied to a tree. Accordingly, as the Varanasi district labour commissioner's statements illustrate, officials tend to conceive of all children, bonded or not, as eligible only for protection from the child labour law.

The Indian government has failed to collect even remotely accurate information on bonded labour. There is no reliable official data for India as a whole on how many bonded labourers there are or what states have done, or failed to do, about it. The statistics that are maintained indicate how little bonded labourers are being protected. The central government relies on data supplied by the state officials-some of the same individuals who told Human Rights Watch that no children were bonded in their territories.

Rehabilitation of Bonded Labourers

Simply freeing a bonded labourer is not enough. For the

same reasons that he or she became bonded in the first place-a lack of land, access to credit, or other forms of employment, or compulsion in the form of caste-based violence and social and economic boycotts from upper castes-without assistance a freed bonded labourer is likely to become bonded again.

According to Kannegi Packianathan, director of the Tamil Nadu State office of the National Commission for Scheduled Castes and Scheduled Tribes, "If they can't get a loan, they abandon it and return to bonded labour." The Indian Supreme Court has recognised that without rapid and effective rehabilitation efforts, freed bonded labourers will quickly relapse into bondage:

> [I]t would be nothing short of cruelty and heartlessness to identify and release bonded labourers merely to throw them at the mercy of the existing social and economic system which denies them even the basic necessities of life. . . . What use are "identification" and "release" to bonded labourers if after attaining their so-called freedom from bondage to a master they are consigned to a life of another bondage, namely, bondage to hunger and starvation where they have nothing to hope for, not even anything to die for. . . . What would they prize more: freedom and liberty with hunger and destitution staring them in the face or some food to satisfy their hunger and the hunger of their near and dear ones, even at the cost of freedom and liberty? The answer is obvious. It is therefore imperative that neither the Government nor the Court should be content with merely securing identification and release of bonded labourers but every effort must be made by them to see that the freed bonded labourers are properly and suitably rehabilitated after identification and release.

Under the Bonded Labour System (Abolition) Act, 1976, bonded labourers have a right to immediate rehabilitation. The Supreme Court has explained that the state's duty to "suitable rehabilitate" its bonded labourers is also required by article 21 of the Indian Constitution, guaranteeing the right to life and liberty, and article 23, prohibiting the practice of debt bondage and other forms of forced labour or slavery. As of 2003, a bonded labourer identified and released by the state is supposed to receive a rehabilitation package of Rs. 20,000 (U.S.$417), paid in equal parts by the Ministry of Labour and the state, that includes:

(i) Allotment of house-site and agricultural land;

(ii) Land development;

(iii) Provision of low cost dwelling units;

(iv) Animal husbandry, dairy, poultry, piggery etc.;

(v) Training for acquiring new skills; developing existing skills;

(vi) Wage employment, enforcement of minimum wages etc.;

(vii) Collection and processing of minor forest projects;

(viii) Supply of essential commodities under targeted public distribution system;

(ix) education for children; and

(x) Protection of civil rights.

Nevertheless, even those districts that manage to free bonded labourers often fail to rehabilitate them. According to staff at an NGO that works to release and educate bonded children: "if [bonded child labourers] get rehabilitated according to the law, they get Rs. 20,000, but the government takes some and by the time it gets to the child, he is an adult." Part of the problem is district magistrates' failure to issue release certificates to bonded labourers when they are freed, without which they are not eligible for rehabilitation. NGO staff in Varanasi explained: "A lot of time is spent getting release certificates.

NGOs spend so much time arguing that it is bonded labour. In Karnataka, of the some 10,000 bonded labourers identified in ten districts following training of government officials on bonded labour in 2000 and 2001, none had been rehabilitated at the time of writing, only some had received release certificates, and of these, none had received the initial Rs. 1,000 due from the government to each bonded labourer upon release. According to Kiran Kamal Prasad, the director of JEEVIKA, the NGO that trained the officials, the released bonded labourers "are in a bad shape. Some have had to migrate to find work." Prosecutions and Convictions Extracting bonded labour is punishable by up to three years in prison and a Rs. 2,000 (U.S.$42) fine under the Bonded Labour System (Abolition) Act, 1976.

However, few employers are prosecuted and even fewer, if any, are convicted. Indeed, Human Rights Watch was not able to learn of a single conviction under the act since 1988. The Ministry of Labour's director general for labour welfare, Manohar Lal, who is responsible for bonded labour, told Human Rights Watch that the central government had no data on prosecutions or convictions under the Bonded Labour System (Abolition) Act, 1976. The Director General said his office has tried, without success, to get this information from the states.

In Tamil Nadu, according to state government officials and documents obtained by Human Rights Watch, only one hundred prosecutions were initiated from 1997 to February 2002, although 37,082 bonded labours were identified during this period. Although Human Rights Watch was not able to obtain data on how many prosecutions, if any, resulted in convictions, a

Tamil Nadu government official who wished to remain anonymous told Human Rights Watch that "most of the cases where charges were started ended up in acquittal . . . ."In Kanchipuram, Tamil Nadu, where there are an estimated 40,000 to 50,000 bonded child labourers in the silk industry, only one bonded labour prosecution had been initiated as of February 2002.

In Varanasi, the deputy labour commissioner told Human Rights Watch that prosecutions had been filed with the court judicial magistrate against the employers in twenty-one bonded labour cases identified in 2001, but none had resulted in conviction as of March 2002. In Bangalore Rural District, Karnataka, no employers have been convicted under the Bonded Labour System (Abolition) Act, 1976, according to a bonded labour caseworker in the office of the district's deputy commissioner; at the time of writing, no one had been convicted in the some 10,000 cases of bonded labourer that the Karnataka state government identified in 2000 and 2001. Human Rights Watch saw evidence that in some areas, the government now is taking steps to train its officials in how to enforce the bonded labour and child labour laws. This needs to be vastly expanded. Training is a critical first step and should be encouraged and conducted nationwide. However, it must also be accompanied by consequences for officials who fail to enforce the laws.

In Tamil Nadu, the government in August 2001 published a manual on enforcing the Bonded Labour System (Abolition) Act, 1976, and rehabilitating bonded labourers. The manual explains what bonded labour is, what the law requires, and how government officials should enforce it. It also clearly states that children are bonded and directs labour inspectors to initiate actions against employers who bond them. Human Rights Watch was told the manual was distributed to every sub-district (taluk). The Kanchipuram district collector independently confirmed that it had been distributed throughout his district, although he is still not implementing the bonded labour law for children in his district, suggesting that further training, monitoring, and accountability is needed.

In Karnataka, following training of officials in ten districts by the NGO JEEVIKA in 2000 and 2001, the state reportedly saw a sharp increase in the numbers of bonded labourers identified in those districts, although Human Rights Watch was not able to learn how many, if any, of these were children. Moreover, at the time of writing, none of these bonded labourers had been rehabilitated. Karnataka's 2001 plan on child labour directs the deputy commissioners (called district magistrates in other states) to use the Bonded Labour System (Abolition) Act, 1976, for children who are bonded. And the state's manual for labour inspectors on prosecuting child labour cases also instructs inspectors to include in their reports "advances/loans taken from employer." However, there is no direction that these cases also be pursued under the bonded labour law; where the child is found to be underpaid, the manual suggests prosecuting under the Minimum Wages Act, 1948, as well as the Child Labour (Prohibition & Regulation) Act, 1986, but does not instruct the inspector that nominal wages can constitute bonded labour. Child Labour Under the child labour regime, the government is still failing millions of Indian children.

Most government officials whom Human Rights Watch interviewed were refusing to fulfill their legal obligation to identify, release, and rehabilitate children in illegal occupations and to fine and imprison their employers. Instead, most favoured persuasive tactics, including public awareness campaigns and voluntary enrollment in transitional schools. While these tactics are not unimportant, for the children who are not free to leave work because they are bound to a master by debt, they are too little too late. States bear primary responsibility for withdrawing children from illegal work, for rehabilitating them, and for punishing employers. The Ministry of Labour is ultimately responsible for the states' compliance with the law and determines which industries are hazardous. The ministry also runs child labour elimination programmes, consisting primarily of transitional schools for some 181,000 former child workers. Because sericulture and silk weaving are classified as hazardous, employing any children in these sectors up to age fourteen-bonded or not-is illegal under Indian law.

Freeing Child Workers

State factory and labour inspectors, National Child Labour Project committees, and in some states, other departments such as the education department, are all responsible for identifying children working in hazardous industries. Although it is not difficult to find children working openly in hazardous occupations, including in the silk industry, in Karnataka, Tamil Nadu, and Uttar Pradesh, as well as other states, these bodies are either not inspecting for illegal child labour at all or are failing to do so effectively. The Supreme Court in 1996 ordered the states to conduct surveys of child labour in their territories, which they did in 1997 and 1998. The reported findings wildly underestimated the incidence of child labour.

For example, Tamil Nadu reported 10,118 children working in hazardous occupations, Uttar Pradesh 32,000 (hazardous and non-hazardous), and Karnataka 44,070 (hazardous and non-hazardous). Government officials continue to rely on these numbers, as well as the 1991 census. Subsequent government efforts to find children working in hazardous occupations do not appear to be more accurate.

Prosecuting Violators

Prosecutions under the Child Labour (Prohibition & Regulation) Act, 1986, for employing children in occupations deemed hazardous have been rare and convictions rarer still. Both appear to have declined since the mid- to late 1990s. The secretary of the Ministry of Labour, Vinod Vaish, and his joint secretary for child labour, K. Chandramouli, refused to give Human Rights Watch nationwide data on prosecutions and convictions under the Child Labour (Prohibition & Regulation) Act, 1986.

Documents obtained by Human Rights Watch, citing data from the Ministry of Labour in New Delhi, stated that in 1996-1997, 13,090 inspections found 509 violations. Of these, 374 were prosecuted and fourteen convictions obtained-a conviction rate of less than 4 percent. Human Rights Watch was unable to verify these data's accuracy. In addition, according to former Secretary of the Ministry of Labour Lakshmidhar Mishra, even where employers are convicted, trial courts impose only negligible fines. Nevertheless, Human Rights Watch was able to obtain statistics at the state level that further reveal how few employers face sanction for using children.

In Karnataka, the government initiated 1,018 prosecutions against employers under the Child Labour (Prohibition & Regulation) Act from 1997 to 2001 and obtained 101 convictions, a conviction rate of less than 10 percent. However, only 986 "child labourers" were "detected" during this same period, suggesting that either the data are flawed or that some prosecutions were for violating other parts of the act (such as conditions of employment where children are working legally). For the 986 child labourers, Rs. 206,975 (U.S.$4,312) in fines was collected from employers-1 percent of the Rs. 20,000 ($417) per child that should have been imposed. Under the Karnataka Shops and Commercial Establishments Act, 1961, which supplements the Child Labour (Prohibition & Regulation) Act, the government identified 1,271 child labourers from 1997-2000, imposed Rs. 58,175 (U.S.$1,212) in fines, initiated 306 prosecutions, and obtained nineteen convictions, a conviction rate of about 6 percent. In Varanasi, according to the deputy labour commissioner in March 2002:

After the Supreme Court's decision, there was good work on child labour. There were about 575 prosecutions under the Child Labour Act from December 1996 until now. About 300 have been decided. One hundred sixty employers have been convicted and given three-month sentences. These were in carpet, the sari industry, brick, automobiles, but not hotel. However, a study commissioned by the National Labour Institute, the Ministry of Labour's autonomous research wing, found that most of these prosecutions-546 of the 575- were launched between August 1997 and March 1999, suggesting that the Varanasi government is no longer seriously pursuing offenders.

In Tamil Nadu, the state identified 10,118 children working in hazardous industries in 1997 and, as of March 2002, had released 157 children and lodged 8,799 criminal cases, most of which ended in acquittal. The decline in prosecutions appears to be in accord with official government policy. In April 2002, high-level officials at the Ministry of Labour told Human Rights Watch that the ministry no longer favoured prosecution under the Child Labour (Prohibition & Regulation) Act, 1986.

Loopholes and Weaknesses in the Child Labour (Prohibition & Regulation) Act, 1986

Even in the uncommon case in which an official tries to enforce the Child Labour (Prohibition & Regulation) Act, 1986, deficiencies and loopholes in the act itself make enforcement very difficult. Compared with the bonded labour law, the child labour law has broad loopholes for family-based labour and determining age, and carries lighter penalties. It is also more difficult for officials to initiate cases.

First, the act excludes any work done in the family or in government-run or recognised schools, even if the act prohibits children from doing that work in any other context. This exception creates an incentive for factory owners to contract with or bond adults for work to be done in their homes. The adults then use their own children or bond other children to help with the work, claiming, if inspected, that the bonded children are their own. This has happened in the silk industry in Varanasi; the match and beedi industries in Tamil Nadu are other well-known examples. Labour inspectors told researchers at the National Labour Institute: "Most of the employers claim the child workers as their family members, while the child workers' physical appearance such as tattered clothes, undernourished and underdeveloped physical physique, etc., belie the employers' claim[s]."

Gilbert Rodrigo, director of the Tamil Nadu-based NGO Legal Resources for Social Action (LRSA), also explained: "Statistically, there has been a decrease in child labour [prohibited by law]. Employers make a contract with adults, then they take the work home and their children do it full time." The project director for Kanchipuram's night schools for working children told Human Rights Watch: "The Child Labour Act is not sufficient-it doesn't provide protection. The Child Labour Act will not help [children] to come out of the labour problem because they are working in homes."

The Child Labour (Prohibition & Regulation) Act, 1986, should be amended so that household enterprises and government schools and training centres are no longer exempted from prohibitions on employing children. Second, unlike the Bonded Labour System (Abolition) Act, 1976, and the Scheduled Castes and Scheduled Tribes (Prevention of Atrocities) Act, 1989, the Child Labour (Prohibition & Regulation) Act, 1986, is not cognizable, meaning that a police officer may not arrest without a warrant. Instead, it is incumbent on labour inspectors to bring cases against employers. Because the act is not cognizable, it is more difficult for inspectors to collect evidence.

Barriers to Law Enforcement

A lack of well-trained staff in sufficient numbers also impedes law enforcement. At both the district and state levels, the staff devoted to enforcing the child labour laws is insufficient. Although some states, including Tamil Nadu and Karnataka, in order to conduct the survey for the Supreme Court, charged inspectors from various government departments, not just the labour department, with finding child labour, these officials have not continued to enforce the law. According to Lakshmidhar Mishra, former secretary of the Ministry of Labour, one reason for the "failure to enforce the plethora of legislative enactments" covering child labour is that a very small number of inspectors are responsible for enforcing a large number of laws in a large number of establishments, which makes it "virtually impossible" for them to do their jobs thoroughly. Many inspectors are not adequately trained to enforce the law. The National Labour Institute has found:

> Most of the inspections conducted on child labour law do not result in prosecutions because, during the inspection, some of the provisions of the Act or the rules are lost sight of by the labour enforcement officers. This may be because of the labour enforcement officers not being careful, thorough and tactful while carrying out inspections. . . . For Inspections under the Act to be purposeful and effective, these should always be surprise inspections. Otherwise, employers of child labour send the children away or hide them and destroy the evidence about their employment the moment they get a hint of the impending visit of the Inspector. . . .

Prosecutions do not result in convictions very often because of the failure on the part of Inspectors to gather sufficient and reliable evidence during the course of the inspections for bringing home the violations of the Act by the accused persons. The National Labour Institute has produced enforcement manuals and training materials for child labour inspectors and is training government officials and others, including district magistrates, inspectors, unions, and NCLP directors. Training is critical and should be encouraged and offered nationwide. However, training officials on how to enforce the law will not get results if government officials at the highest levels discourage law enforcement; if problems of corruption, apathy, and caste and class bias are not addressed; and if the failure to enforce is not sanctioned-or even monitored.

Long delays and low conviction rates also discourage officials from pursuing prosecutions. It should be noted, however, that some of the same officials who complain about the process bear responsibility for improving, not abandoning, it. The district collector of Kanchipuram, who is responsible for the administration of justice, told Human Rights Watch that his office does not support prosecuting employers. Instead, it is trying to address the lack of education by providing night schools and to address poverty by organising women's self-help groups.

National Child Labour Policy (NCLP) and the National Authority for the Elimination of Child Labour

The government's primary strategy for child labour has been to run schools designed to lure children out of hazardous labour and prepare them for formal schools or for a vocation. These schools are created through the National Child Labour Policy (NCLP), which was announced in 1987 and expanded in 1994 by the National Authority for the Elimination of Child Labour. The programme operated in one hundred districts in thirteen states in 2002. In 2001-2002, Rs. 670 million (U.S.$14 million) was allocated for the NCLP programme, almost double that from the previous year. The policy is administered by committees (entirely distinct from the bonded labour vigilance committees).

The NCLP committees are charged with generating awareness about child labour, conducting surveys to identify child labour, running special schools that provide vocational and non-formal education and mainstream children into formal schools, enforcing child labour laws, and conducting adult education and activities related to income and employment generation. In practice, most committees just run the special schools. According to the government's own evaluation in 2001, the committees do almost none of the other tasks, and many hardly meet at all. The evaluation concluded that "[t]he enforcement of child labour laws has seen virtually no improvement ever since the projects were launched, with one or two notable exceptions."

A separate evaluation in 2000 of NCLP projects in Andhra Pradesh, Tamil Nadu, and Karnataka found similar results in those states. Human Rights Watch's investigation in Varanasi suggests that the Varanasi NCLP committee has done slightly better. In addition to supervising thirty-six schools, the committee is conducting surveys of child labour. In contrast, in Karnataka, local NGOs said that most committees were defunct. According to a local activist: "We called a meeting of the committee [in Ramanagaram], and the members didn't know each other. . . . The only reason they meet in Ramanagaram is because we created it. The other committees haven't met." Another frequent criticism is that the NCLP committees, like the bonded labour vigilance committees, are staffed by those apathetic to or actively employing child labour.

There is a conflict of interest-members of the committees often work in industries that employ child labour," the activist said." Many members of the committee don't even believe in the eradication of child labour. "Operating special schools for former working children who lag far behind children their age in regular schools and may not be acclimatised to a school environment is, in theory, a good idea. However, the NCLP schools serve very few children compared with the number still working, and channel even fewer into a vocation or into formal schools.

Most of the 3,369 schools operating in 2001-2002 opened in the preceding two years and covered 181,004 children in one hundred of India's 593 districts. This was an increase of only 792 children from the previous year, and only 1.61 percent of the 11.29 million working children recorded by the 1991 census (and a much smaller percentage of the tens of millions of children actually working). The Ministry of Labour has admitted that the "number of working children who have been covered by the special schools is a miniscule of the total number of working children awaiting to be released and rehabilitated." Moreover, not all of the 181,004 children in NCLP schools are former child labourers.

NCLP schools promise students a hot meal and a Rs. 100 (U.S.$2.08) stipend not offered in regular schools, and some operate in areas without a formal school nearby. Understandably, they attract non-working children, some of whom come from formal schools. The head of an NGO running several NCLP schools for the government told Human Rights Watch: "The [NCLP] schools are good for children who are working within the family and in the community, but the Supreme Court intended for them to be for children released from hazardous industries, and these are not the children who attend the schools."

A 2000 study of NCLP schools in Andhra Pradesh, Tamil Nadu, and Karnataka confirmed that the schools attract non-working children, adding specifically that the schools had not reached working children "who are in the strong clutches of money lenders, hazardous industries, [and] labour contract" because they were more difficult to get released. Instead, the schools targeted "only those in the un-organised sector and school dropouts" because they were more easily found.

The government's 2001 evaluation of the NCLP found that where schools regularly delivered the monthly stipend and hot meal, they were able to hold students.

However, only 38 percent of surveyed schools paid stipends on a monthly basis; for the rest, there were delays of many months, which caused children to drop out of school. In Bangalore Rural District, Karnataka, which was not part of the evaluation because the first NCLP schools began functioning in July and August 2001, 1,620 children had been admitted as of February 5, 2002. However, attendance rates were low (averaging 62.5 percent), infrastructure poor, vocational education inadequate or nonexistent, and mainstreaming into formal school unsuccessful.

A study of NCLP schools in Andhra Pradesh, Tamil Nadu, and Karnataka found that children often returned to work after leaving the NCLP schools, a fact partially confirmed by the project director for NCLP schools in Bangalore Rural District. Thus, the NCLP schools have presumably successfully mainstreamed even fewer than the 133,000 children that the government claims-not even keeping pace with the rate of new children entering hazardous work, much less decreasing the number who are already there.

Non-governmental Programmes

In addition to and, in some cases, in cooperation with the Indian government, a great variety of domestic and international organisations, including UNICEF, CARE, the U.S. Department of Labour, and the International Labour Organisation through its International Programme on the Elimination of Child Labour (IPEC), have programmes related to reducing child labour. These programmes tend to focus on running schools for former working children, raising public awareness that child labour is wrong, and lobbying government officials to enforce and strengthen the child labour and education laws.

Both activists and government officials singled out the M.V. Foundation's work in Andhra Pradesh as particularly successful. These projects contribute to the rehabilitation of child labourers, and, in many places, are the only forms of rehabilitation operating. However, as explained below, unless the Indian government enforces its child and bonded labour laws and removes and protects bonded children, the projects often miss bonded children, who are not free to leave their employers voluntarily.

Rehabilitation-Cum-Welfare Funds

Rehabilitating child workers is critical to preventing them from immediately returning to work. It is generally conceived of as providing education to the child and employment or sources of credit to parents. Rehabilitation of child labourers became legally required in 1996, when the Supreme Court, finding continued use of child labour prohibited by the Child Labour (Prohibition & Regulation) Act, 1986, ordered state labour inspectors to identify children employed illegally, to fine employers Rs. 20,000 (U.S.$417) per child, and to deposit this money in a "Child Labour Rehabilitation-cum-Welfare Fund," the income on which must be used only for the child concerned.

Where the child is employed in a factory, mine, or other hazardous work, the Court requires the state to employ an adult family member in lieu of the child or, where this is not possible, to deposit an additional Rs. 5,000 (U.S.$104) into the fund on the condition that the child be withdrawn from work and sent to school. Parents should then receive the income on the Rs. 25,000 (U.S.$521) for as long as they continue to send the child to school. The Court ordered the Ministry of Labour to monitor the implementation of its decision.

Despite the Court's 1996 order, by 2003, states had collected little money from employers. Even where they had, the money was not used to rehabilitate children. The Ministry of Labour's joint secretary for child labour refused to give Human Rights Watch nationwide data on monies in the rehabilitation-cumwelfare funds. However, he did say that the money collected in the past was "meager" and that very little money was currently being collected because there was little use of child labour in hazardous occupations.

National Human Rights Commission's Role

By focusing on bonded and child labour, the National Human Rights Commission (NHRC) has changed the environment in the areas where it has investigated. However, while it can encourage law enforcement, with limited powers and resources it cannot replace it. The NHRC is the government of India's autonomous human rights body. Created pursuant to the Protection of Human Rights Act, 1993, its statutory functions include conducting independent investigations into human rights abuses, both on its own initiative and in response to petitioned complaints.

In conducting these investigations, the commission enjoys the same powers as a civil court and can, accordingly, summon witnesses and examine them under oath; compel discovery and production of documents; and requisition public records from a court or office. However, unlike a civil court's ruling, the NHRC's recommendations are not binding, although the commission has been able to pressure the government into complying with its recommendations on many occasions. The commission has made general interventions on child labour, especially in Uttar Pradesh but also in the south, and intervened in some individual cases. Following the Supreme Court's request in 1997 that it monitor states' implementation of the bonded labour law, it appointed various special rapporteurs to address the issue.

The NHRC makes a huge difference, especially in the quality of the NCLP programmes. They conduct field visits to monitor-they look at education, health and nutrition, the stipends, and the resources for rehabilitation. At least this conversation is happening so people sleeping on the job wake up. The NHRC recognises the job of good people and goes against people who are not working.

14

Sericulture in Rural Areas

Sericulture is an agro-based labour intensive industry which provides gainful employment to the rural and unemployed youth and helps to uplift the socioeconomic status of small and marginal farmers . The economic advantages of Sericulture industry lies in its high employment potential with low investment. One hectare of mulberry creates employment to persons throughout the year. 60% of them are women thereby supporting a greater role for women in development. It requires low gestation period and continues to yield for 15-16 years with little expenditure on maintenance. It gives higher returns, unlike other agricultural crops.

Andhra Pradesh occupies 1st position in productivity and 2nd position in the country next to Karnataka in production of Silk. Andhra Pradesh produces all the four popular varieties of Silk worm cocoons namely Mulberry, Tasar, Eri and Muga. Andhra Pradesh has got very strong and traditional weaving base with more than a lakh number of hand looms mostly concentrated in weaving pockets like Dharmavaram, Pochampally,, Gadval, Patur, Peddapuram,Narayanpet, etc., In the last decade number of cotton weavers have taken to silk weaving in centers like Rayadurg and Proddatur because of better income in silk weaving. As a rural agro based industry, Sericulture has now expanded to almost all districts in the state.

Sericulture at a glance

S.no	Item	Units	2005-06
1	Employment generation	Lakh persons	4.88
2	Area under mulberry cultivation	Cumulative in acres	97,646
3a	CB Cocoon production	In MTs	46,785
b	Bivoltine cocoon production	In MTs	1239
	Total Cocoon Production	In MTs	48024
4	Cocoon production / 100	Dfls In Kgs	54.4
5	Raw silk production	In MTs	5,336
6	Sericulture farmers	In Nos	76,971
7	Tasar food plantation Available for rearing	In acres	25000
8	Tasar Dfls brushed	Lakh nos.	8.82
9	Tasar Cocoon production	Lakh nos.	113.53
10	Tasar rearers	(In No.s).	3,000
11	Area under Castor& Topioca	In lakh Hectares	3.74
12	Eri Dfls brushed	(In No.s).	1.759
13	Eri cocoon production	MTs	33.500

Objectives of the Department of Sericulture During 2006-07

— To stabilise the existing area under mulberry and bring an additional area of 8700 acres under mulberry cultivation with High Yielding Varieties.

— To produce 51415 tonnes of reeling cocoons.

— To produce and supply quality Cross Breed Disease Free layings through Departmental Grainages.

— To increase productivity at the farmer level to 55 Kgs per 100 dfls.

- To generate employment to 4.35 lakh persons both directly and indirectly.
- To produce and supply quality Tasar layings to 3,000 rearers who are mostly Tribals.
- To produce 2400 M.Tonnes of bivoltine cocoons for the production of Globally competitive Internationally graded Silk required to face the challenges of W.T.O.
- To improve productivity levels in Tasar culture
- Expansion of Ericulture to the potential districts of Telangana and Coastal areas where castor and tapioca are abundantly grown .Ericulture is also being introduced on pilot basis in perennially drought affected district of Anantapur.
- To implement Economical water management system through Drip Irrigation.
- To disseminate Scientific Sericultural Practices by promoting modernisation and quality up gradation of the industry.
- To develop of Region specific & season specific races through APSSRDI, Hindupur.
- To integrate of Non Farm Activities ie. Reeling and Twisting with weaving and to strengthen linkages between the farm and non farm activities.

Strategies of the Department

Though the present production of cocoons per unit area is more remunerative when compared to the other commercial crops, this may not help for the sustenance of the industry in the long run. Therefore, it is imperative to increase the productivity levels on par with sericulturally advanced countries. The department has identified certain gaps, which are coming in the way of producing qualitative and quantitative production of cocoons.

On-farm Sector

- — Expansion of mulberry with High Yielding Varieties.
- — Maintenance of Separate Chawkie Gardens(for young age silk worm rearing to reduce mortality).
- — Effective Water Management like "Drip Irrigation".
- — Appropriate Disinfection Technology.
- — Separate pucca rearing house.
- — Shoot harvesting and Shelf rearing.
- — Rotary Mountages.
- — Human Resource Development.
- — Research & Development.
- — Special thrust for promotion of Bivoltine Hybrid cocoons & Production of International grade silk.

Non-farm Sector

Inspite of all out efforts being made to modernise the industry the post cocoon activities especially reeling activity is not emerging as economically viable due to high investment with low profit margin. As a result, most of the cocoons produced in the State are making way to the neighboring states. The strategies adopted for non-farm sector development are:

- — Working capital assistance to the non-farm sector units.
- — Entrepreneur Development Programs
- — Assistance for establishing silk reeling, twisting, looms, processing units.

— Review of silk import policy
— Setting up of improved multiend reeling machines
— Providing of interest subsidy on loans & working capital for the reelers having cottage and Multiend basins to reduce the interest burden to the entrepreneurs.
— Computerised linking of all the Cocoon Markets in the State to know the information on Market rates both for rearers & reelers The Department has planned to revive 204 Silk Reeling basins with the annual production capacity of 42 MTs from the existing level of 70 basins with 14 M.tonnes. It is also planned to establish the required number of basins and spindles to consume the entire quantity of cocoons that would be produced in a phased manner.

Departmental Infrastructure

The available technical infrastructure details is as follows:

S.No.	Name of the Unit	Number of Units
01	On Farm Technical Service Centers	140
02	Seed farms	39
03	Seed area TSCs	9
04	Seed Cocoon Markets	4
05	Grainages	13
06	Govt. Cocoon Markets	8
07	Reeling units	15
08	Non-farm TSCs	11
09	Training Centres / Units	8
10	Tasar seed stations	9

Progress During 2005-06

— A quantity of 1239 MTs of Bivoltine cocoons was produced for the first time in the state by dovetailing all associated facilities like rearing sheds, rearing equipment and drip irrigation system etc., to the Bivoltine farmers.

— 1595 separate rearing sheds have been constructed. So far a total of 10099 separate sheds were constructed under Catalytic Development programme for the benefit of farmers and to increase bivoltine sericulture.

— An extent of 1462 acres has been covered under the Drip Irrigation programme .

— An extent of 7834 acres of new area has been covered under mulberry.

— Bivoltine Silk Worm eggs supplied to Bivoltine farmers on subsidised cost @ Rs. 100/- per 100 Disease free layings _ 3,000 tribal families are engaged in rearing of Tasar silk worms on the food plants available in the forest areas of Adilabad, Karimnagar, Warangal and Khammam Districts and produced a quantity of 113.535 Lakh number of tasar cocoons.

— The Department in co-ordination with the Central Silk Board under Catalytical Developmental Programme is implementing an Eri Project to create an additional income to Castor/Topioca growers through rearing of Eri silkworms in Nalgonda, Mahabubnagar and East Godavari districts.

— A quantity of 33.50 Metric tonnes of Eri cocoons was produced by 811 Castor and Tapioca farmers which is a record. Ericulture is providing additional income to farmers.

— Special programmes such as Mass Disinfection Programme, Training programme etc., are being implemented for the benefit of small, marginal farmers, Reelers, Scheduled Castes, Scheduled Tribes and Women to enable them to increase the productivity and their economic status.

— To encourage the Private reelers and to strengthen Non Farm Sector, schemes like Production incentive to Multiend reeling units on Production of Quality Silk Yarn, Interest subsidy on loans and working capital for the reeling units sanctioned by Commercial Banks are being implemented Production incentive to 21946 Kgs of Multivoltine and 12974 Kgs of Bivoltine Silk has been provided to the reelers.

— Providing assistance to the drought affected farmers for revival of mulberry.

New Initiatives of the Department

— 16 Multiend reeling units were revived under private sector. In addition to this 15 Govt. Silk Reeling units were also revived in the State for effective utilisation of the services of the workers and to help the farmers in selling their cocoons at competitive prices.

— To encourage post cocoon activity and to bridge the gap between the on farm and non farm sectors to maximum extent, it is proposed to establish one Chinese automatic reeling unit under private sector.

— To provide marketing facility to the ERI farmers, it is proposed to establish one ERI processing cum spinning unit under private sector.

— Expansion of Tasar and Ericulture to other potential Districts in the State.

Focussed Areas:

Seed production

A systematic 3 tier system of seed organisation programme has been established i.e., at P3, P2 and P1 level multiplication for both local and foreign races, to supply quality seed cocoons for production and supply of disease free silkworm eggs to the farmers. The State has achieved self sufficiency in the production of seed cocoons for the production of disease free silkworm eggs in Government grainages. Adequate infrastructure with supporting staff is available for smooth functioning of seed organisation.

Extension support

The Department is advocating new technologies in both Mulberry cultivation and silkworm rearing to the farmers through Technical Service Centres in order to increase cocoon production in the State and to increase productivity per acre. Study tours are regularly organised and extension literature on latest techniques is distributed regularly. Mobile Farmers Training Units are also involving in transfer of new technologies to all Sericulture villages. Average productivity per 100 DFLs in several areas of the State has reached 54.40 Kgs.

High Yielding Variety Mulberry Plantation

The main focus will be to increase the productivity and thereby the income per unit area. In order to increase the productivity of leaf yield, major thrust has been given to increase the area under high yielding varieties of mulberry saplings with wider spacing. The farmers have realised the importance of quality consciousness of mulberry leaf varieties and are voluntarily coming forward to take up V1 & S-series variety of plantation through saplings with wider spacing and also paired row system. The farmers are replacing the existing mulberry varieties with V1& S-series in a phased manner.

Development of Kissan Nurseries

Taking into consideration of the growing demand for V1 Mulberry Plantations, the farmers are being encouraged to take up Kissan nurseries on commercial lines to meet the demand of the farmers for saplings. It is already proved as a viable commercial proposition and the farmers are raising nurseries twice in a year during February and July. Plantation through Saplings have many advantages over plantation through cuttings Advantages of saplings:

- — High survival rate due to existing root system.
- — Higher area coverage due to better use of plant material.
- — Scope for removal of undesirable variety at nursery stage.
- — Quick and better establishment of garden.
- — Plants can be trained suitably.
- — Uniform development of the garden.
- — Saplings can be raised in time to take up timely plantation during monsoon.
- — Minimum of crops can be taken during the 1st year of establishment.

Development of Drip Irrigation

As a water conservative measure, much impetus is being given for Drip irrigation which is proved as most effective in economic use of water for irrigation and is well suited for mulberry plants. The financial assistance available to the farmers through the Centrally Sponsored CDP Scheme of CSB is effectively utilised for the promotion of large scale adoption of this technology at the ground level.

Adoption of more crop cycles per year:

The farmers are dividing their mulberry gardens in to two plots and are harvesting 8-10 crop cycles per annum. By this the farmers are able to get income almost every month thereby improving their economic status.

Development of Bivoltine Rearing Technology

Realising the importance of quality and productivity for sustenance of sericulture industry, the department has formulated a massive Bivoltine Programme in the State by identifying nucleus villages and propagating Bivoltine Technology. This could be achieved only by providing incentives to the Bivoltine farmers for construction of Rearing Sheds, Rearing Equipments, Drip irrigation facility, supply of disinfectants etc,. By creating the above associated facilities, the farmers are able to improve their productivity and are producing quality Bivoltine Silk of International Grade.

Biovoltine Hybrid Programme

The Department is implementing Bivoltine hybrid programme throughout Andhra Pradesh with the financial assistance of Central Silk Board & Government of Andhra Pradesh. A record production of 1239 M.T of bivotine cocoons was achieved during 2005-06

Promotion of Chawkie Rearing Centres (CRCs)

The farmers have conceived the concept of supply of

chawkie worms as a commercial proposition in view of the great acceptance for rearing of worms supplied through chawkie which reduces the period of rearing at farmers level with assured increase in productivity. The farmers are coming forward to run chawkie rearing centres in private sector voluntarily. The financial assistance available under CDP is being effectively utilised by such farmers who are willing to construct chawkie rearing buildings and purchase the chawkie rearing equipments.

Marketing Faciliites

Marketing facility to the cocoons produced by the farmers is provided through (8) regulated Cocoon Markets and (16) Notified Markets. To ensure a reasonable price to the farmers, cocoon testing procedure was introduced in all the Cocoon Markets which is found to be more beneficial to the farmers. This will induce quality consciousness among farmers and reelers. For the benefit of farmers and reelers, Information Centers have been established in the Cocoon Markets to depict various improved technologies mainly through audiovisual Aids. Cocoon Market rates are being broadcasted daily through All India Radio and popular T.V channels.

Silk Reeling

Though Sericulture has been developing very rapidly in respect of cocoon production in the State, reeling activities are yet to be developed at the same pace. There is still a wide gap between the production of cocoon and consumption with in the State. Non-Farm Technical Service Centers have been set up for the exclusive promotion of post cocoon activity as there is good potential for the development of Non-Farm sector domestically.

Promoting reeling industry

1. An incentive of Rs.100/- per kg. of silk is being given to the Reelers under CDP programme to encourage Bivoltine Silk Production on

Multiend silk reeling units

2. An incentive of Rs.50/- per kg for the production of Multivoltine Silk reeled on Multiend Silk Reeling Units which is suitable for warp is also being given to reelers under SDF.
3. Department is providing Skill up-gradation training programme to the workers of Improved Multi-end Silk Reeling units on stipendiary basis including 25% cost of raw material .
4. Subsidy to the maximum of 50% on working capital loan will be paid to the financial institutions by the Department on behalf of silk reelers to minimise the burden on small and marginal silk reelers to encourage and also for sustenance of non-farm units.
5. 50% Subsidy on machinery cost is proposed to be provided besides 100% subsidy on CST and Transportation charges, where the traditional charka silk reeling basins are being replaced with Improved Cottage Basin Silk Reeling basins.
6. Assistance is being extended for establishment of Improved Multi-end Silk Reeling units (15:35:50) by DoS-AP, CSB and beneficiary respectively on the unit cost of 6, 10 and 20 basin Improved Multi-end Silk Reeling Units.

7. Special package of scheme is also being implemented for payment of Rs.75/- per kg of Multi-voltine and (Yellow) Rs.125/- per kg of Bivoltine (White) silk as production incentive in non-traditional districts of Hyderabad and E.Godavari / W.Godavari / Krishna as "Incentive for production of quality raw silk on Improved Multi-end Silk reeling Units."
8. 50% subsidy is being given to the progressive multi-end silk reelers on the cost of plant/ machinery to establish reeling accessories like pressurised cocoon cooking system / water softeners / boilers for production of quality raw silk under the scheme "Assistance for purchase of accessories for Improved Multi-end silk reeling machinery".
9. For consumption of defective reeling cocoons and to encourage small silk reelers, who can not afford for Improved multi-end silk reeling machinery, which is more expensive, necessary assistance to establish Improved Cottage Basin / Motorised Charka / Dupion Charka reeling machinery is being provided under CDP.

Tassar Sector

An extent of 8400 hectares of Tasar food plants are available for conduct of Tasar silk worm rearing in the forest. 3000 Nos. of Tasar rearers who are mostly tribals are engaged in Tasar silk Worm rearing. During 2005-06 a quantity of 113.535 lakh numbers of Tasar cocoons was produced by implementation of CDP schemes under Tasar sector.

Ericulture Project

The Department is implementing Eri Project to create additional income of Rs 3000/- to Rs 4000/- per acre to Castor / Topioca growers through rearing of Eri Silkworms. The Project is being implemented on pilot basis in Nalgonda, Mahabubnagar and East Godavari districts in co-ordination with the Central Silk Board . A quantity of 33.50 tones of Eri cocoons was produced during 2005-06. It is planned to expand the programme to Ranga Reddy, Medak, Nizamabad, Prakasam, Srikakulam, Vizianagaram & Visakhapatnam Districts.

Research and Development

Andhra Pradesh State Sericulture Research & Development Institute has been set up as an autonomous organisation in Andhra Pradesh at Kerikera village near Hindupur in Anantapur District to cater to the research needs of the State. The main objective of this institute is to carry out research on all aspects of Sericulture with a special emphasis on adaptive research and to develop region specific, and season specific silkworm breeds. The Research work on maintenance of breeding stock, purification and isolation of inbreeding are under progress. Testing of several Bi-voltine hybrid combinations at different regions is under progress.

A new Multi X Bi-voltine hybrid namely "SWARNANDHRA" has been successfully released by AP State Sericulture Research & Development Institute, Hindupur. Similarly two new Bivoltine hybrids, namely "HEMAVATHI" and "KALPATHARUVU" have also been released and these races have resulted in increased output of white cocoons, which have been found to be suitable to produce silk yarn of international grade.

Andhra Pradesh Silkworm Seed and Cocoon (Control) Act 1956

Section - 1

Short Title:

Sub Section 1: This Act may be called the Andhra Pradesh Silkworm Seed and Cocoon (Control) Act 1956.

Sub Section 2: It extends to the whole of the State of Andhra Pradesh

Sub Section 3: It shall come into force on such date as the Government may, by notification in the Andhra Pradesh Gazette appoint.

Section 2: Definitions:-

In this act, unless there is anything clause repugnant in the subject or context-

aaa) Director means the Director of Sericulture, Andhra Pradesh

aa) Cocoon Market means a market established under Section 5(a) for the sale or purchase of cocoons of all kinds intended for reeling

a) Cocoon means cocoons produced by mulberry or Tasar silkworms either green or stifled, dried, or in any other state or condition but does not include pierced cocoons.

Explanation: in this clause pierced cocoons means a cocoon from which moth has cut out.

b) Government means the State Government

c) Licence means a licence granted under the Act

d) Licencing authority means the head of the Sericulture as section in the Dept. of Ind and Com.,

Andhra and includes any other officer appointed by the government by notification in the Andhra Pradesh Gazette to exercise all or any of the powers or to perform all or any of the duties of a licencing authority under this Act.

dd) Licenced buyer means a person who is licenced to purchase cocoons in a Cocoon Market

e) "Prescribed" means prescribed by rules made under this Act.

f) "Rearer" means a person engaged in rearing of silkworms for the production of silkworm cocoons, whether for reproduction or reeling.

ff) "Rearer" includes all operations from the incubation of silkworm eggs and brushing of silkworms to the harvesting of cocoons.

g) "Silkworm" includes mulberry silkworm, Tasar Silkworm, Muga Silkworm and Eri Silkworm.

h) "Silkworm Seed" means silkworm cocoons, moths, eggs or young silkworms of whatever description intended to be used for the purposes of re-production or rearing.

Section 3. Regulation of Manufacture ETC of Silk-worm Seed:

No person shall manufacture, store transport, sell or otherwise distribute silkworm seed except and in accordance with the terms and conditions of a licence issued under this Act.

Section 4: Regulation of Rearing

No person shall rear silkworm from silkworm seed other than silkworm seed obtained from a person who holds a licnece under this Act.

4-A.: Regulation of Disposal of Silkworm Cocoons

No person shall dispose off or agree to dispose off or pursuance of an agreement entered into make delivery of silkworm cocoons except to persons holding a licence under this Act.

4-B: Regulation of Sale or Purchase of Silkworm Cocoons

1) In any areas in which a Cocoon Market is established under this Act.
   a) No rearer shall sell or agree to sell
   b) No person shall purchase or agree to purchase silkworm cocoons except in such Market and except in such manner as may be prescribed.
2) Where a Cocoon Market is established for any area no person shall, except in such Cocoon Market use or permit the use or assist in the use of, any building room, tent, enclosure, vehicle vessel or place in such area for the sale or purchase of silkworm cocoons, or in any manner aid or abet the sale or purchase of silkworm cocoons.

4-C: Regulation of Transport of Silkworm Cocoons

1). No person shall transport out of the State, silkworm cocoons except under a permit issued by the Director or any other Officer of the Sericulture Department authorized by him in this behalf, and on such terms and conditions and on payment of such fees as may be prescribed

2). Any permit issued under Sub-section (1) may be cancelled by the Officer aforesaid for the breach of any of the terms and conditions subject to which it was issued or for any other reasons to be recorded in writing.

Section 5: Application for Licence

Every application for the grant of a licence under this Act shall be made to the licnecing authority in such form as may be prescribed.

5-A) Establishment of Cocoon Markets

1) The Government may, from time to time by notification specify the places at which Cocoon Markets, Cocoon Market yards and Cocoon stores shall be located

   b) Appoint a market Officer, and constitute a Committee consisting of two representatives of rearers and one each of licenced buyers incharge of charka establishments and filature establishments, with the Market Officer as Chairman for regulating, generally and procedure for the conduct of business in the Cocoon Market and for the performance of such functions as may be determined by the Government.

   Provided that the Director, or any Officer of the Sericulture Department authorized by him in this behalf shall perform the functions of the Market Officer, until a Market Officer is appointed and of the Committee until a Committee is constituted, under this clause.

   c) Specify the Sericultural areas in the State to be served by such Cocoon Market, where silkworm cocoon produced within such areas shall be sold.

d) Assign Markets in which any licenced buyer may carry on his business.

e) All transactions involving the sale or purchase of cocoons in a Cocoon Market shall be by open auction, the payment of the price shall be in cash and the cocoons shall be sold by weight, as required by or under the Andhra Pradesh Weights and Measures (Enforcement) Act 1958.

Section 6: Power of Inspection

1. The licencing authority or any officer authorized by the licencing aurhority in writing on his behalf may with a view to securing compliance with this Act or with any rules or otder made there under

   a) Require any person licenced under this Act ot furnish such information as may be specified

   b) Inspect or cause to be inspected any return, document or account book in the possession of a person, in licneced under this Act.

   c) At all reasonable times enter and inspect any land, building, vessel, vehicle or place or a person licenced under this Act.

   d) At all reasonable times enter and search any land, building, vessel, vehicle or place wherein or in any part of which silkworm seed or cocoon is stored or is being transported and seized or authorize any person to seize any silkworm seed or cocoon including any vessel, receptacle, apparatus, package or covering in which such seed or cocoon is contained if the licencing authority or the knowledge or from information given by any person and taken down by him in writing that contravention of this Act or any rule or order made there under has been or is being committed.

Section 6: Subsection - 2:

Every owner, occupier or other person incharge of any such land, building, vessel, vehicle or place shall be bound to give all reasonable facilities to the licencing authority or the Officer aforesaid in carrying out his inspection under Sub-section.

Section 7

Sub Section-1:

Penalities: Any person who contravenes or fails to comply with all any of the provisions contained in this Act or any rule or ordered made there under shall be punishable with fine which may extend to two hundred and fifty rupees any court trying any such contravention may or failure may direct that any silkworm seed or cocoon or any vessel, receptacle, apparatus, package or covering containing the same in respect of which the court is satisfied that such contravention or failure has occurred shall be forfeited to Government.

Sub-section-2: Any person who contrivance the provisions of clause (b) of sub section 1 of section 4 (b) or sub section 2 thereof, shall be punishable with imprisonment which may extend to 3 months or with fine which shall not be less than five hundred rupees but which shall not exceed one thousand rupees or both.

Section 7 (A) : Compounding of offences:

(1) The Director of such officer as may be prescribed may

(a) Accept, way of composition, from any person who committed or in respect of whom it can be reasonably inferred that he has committed, any offence under the provisions of this Act or the rules made there under, such sum of money not exceeding the amount of fine payable for such offences, as may be determined by the Director or such officer.

(b) In all cases in which any property has been seized as liable to confiscation under this Act release the same on payment of the value thereof as estimated by the Director or such officer.

(2) On payment by such person the sum of money or the value or both as the case may be, the person, if in custody, shall be set at liberty, and all the property seized may be released and no proceedings shall be instituted against such person in any criminal court. If proceedings in any criminal court have been instituted against such person in respect of the offence the acceptance of composition shall be deemed to amount to an acquittal and in no case, any further proceedings shall be taken against such person or property, with reference to the same act or in respect of the same offence.

Section 8: Previous Sanction for Prosecution

No prosecution shall be instituted under section 7 without the previous sanction of the Director.

Section 9: Offences By Corporations etc.

Where a person committing any offence punishable under this Act or any rule or order made there under is a company or an association or body of persons, whether incorporated or not, every director, manager, Secretary, agent or other officer or person concerned with the management thereof , shall, unless, he proves that the offence was committed without his knowledge or that he has exercised all due diligence to prevent its commission, be deemed to be guilty of such offence.

1) Offences by Companies: Where an offence under this Act has been committed by a Company, every person who, at the time of offence was committed was incharge of, and was responsible to, the company for the conduct of the business of the company, as well as the company, shall be deemed to be guilty of the offence and shall be liable to be proceeded against and punished accordingly:- Provided that nothing in this sub-section render any such person liable to any punishment if he proved that the offence was committed without his knowledge or that he exercised all due diligence to prevent the commission of such offence.
2) Notwithstanding anything in sub-section (1) where an offence under this Act has been committed by a company and it is proved that the offence has been committed with the consent or connivance of or is attributable to any neglect on the part of any Director, Manager, Secretary or other officer of the Company such Director, Manager, Secretary or other officer of the Company shall also be deemed to be guilty of that offence and shall be liable to be proceeded against and punished accordingly.

Explanation: For the purpose of this Section:-

a) "Company" means any body corporate and includes a firm or other association of individuals and

b) "Director" in relation to a firm, means a partner in the Firm.

Section- 10: Inquirty and Trial of Offences

No offence punishable under this Act or any rule or order made thereunder shall be inquired into or tried by any court inferior to that of a Magistrate of the Government.

Section-10-(A): Suspension or Cancellation of Licence

1. The Director, if satisfied that any person, holding a Licence under this Act has contravened the provisions of this Act. or the rules made there under or any other terms and conditions of the Licence, may, without prejudice to any punishment under Section 7 after giving the person concerned an opportunity of being heard, suspend or cancel the Licence granted to such persons.
2. Any person aggrieved by the order of suspension or cancellation of licences may appeal to to the Government within such time and in such manner as may be prescribed, and the decision of the Government on such appeal shall be final.

Section 11: Protection of Government Officers and Servants:

1. No suit, prosecution or other proceeding shall lie against any officer or servant of the Government for any act done or proposing to be done under this Act without the previous sanction of the Government.
2. No Officer or servant of the Government shall be liable in respect of any such act in any civil or Criminal proceeding if the act was done in good faith in the course of the execution of duties or the

discharge of functions imposed by or under this Act.

Section - 12: Exemption:

Nothing contained in this Act shall apply to any place where silkworm are reared by any department of Government.

1) The Government may, by Notification in the Andhra Pradesh Gazette, make rules to carry out the purposes of this Act.
2) In particular and without prejudice to the generality of the foregoing power, such rules may provide for
   - a) The form and manner in which applications for licences may be made and the fees for the grant of such licences.
   - b) The terms and conditions which may be included in any licence.
   - bb) The grant of duplicate licences and the renewal of licences and the fees for the same.
   - c) Appeals from any order under this Act, the authority to whom such appeals shall lie, the time within which such appeals shall be made and the procedures for dealing with such appeals.
   - cc) The manner of constituting the Cocoon Market Committees and the powers functions and duties of such Committees and the Market Officer and the sitting fee and other allowance payable to the members of the committees.
   - d) The disposal of the silkworm seed or cocoon or the vessel receptacle apparatus, package or

covering containing the same which may be noticed under Section 6.

e) The Market fee payable by the rearers and the licenced buyers in respect of cocoons sold and purchased in the Cocoon Market, such fee not exceeding two percent of the price amount of the cocoons, and the purposes for which the fees may be utilized.

3) Every rule made under this Act shall, immediately after it is made, be laid before each House of the State Legislature if it is in session and it is not in session in the session immediately following, for a total period of fourteen days which may be comprised in one session or in two successive sessions, and if, before the expiration of the session in which it is so laid or the session immediately following, both houses agree in making any modification in the rule or in the annulment of the rule, the rule shall from the date on which the modification or annulment is notified, have effect only in such modified form or shall stand annulled, as the case may be, so however, that any such modification or annulment shall be without prejudice by the validity of anything previously done under that rule.

Bibliography

Barnard, Charles, *Silk Culture at Home.*

Christian, David "Silk Roads or Steppe Roads? The Silk Roads in World History." *Journal of World History* 11:1 2000.

Di Cosmo, Nicola, *Ancient China and Its Enemies: the Rise of Nomadic Power in East Asian History*, Cambridge, UK; Cambridge University Press, 2002.

Elisseeff, Vadime (Editor), *The Silk Roads: Highways of Culture and Commerce*, Berghahn Books, 2000.

Fischer, H. W., *Silk culture.*

Foltz, Richard. *Religions of the Silk Road: Overland Trade and Cultural Exchange from Antiquity to the Fifteenth Century*, New York: St. Martin's Press, 1999.

Good, Irene, "On the question of silk in pre-Han Eurasia" *Antiquity*, Vol. 69, Number 266, December 1995.

Hill, John E., *The Peoples of the West from the Weilue: A Third Century Chinese Account Composed between 239 and 265 CE*, Draft annotated English translation, 2004.

______________., *The Western Regions according to the Hou Hanshu, Draft annotated English translation*, 2004.

Kuhn, Dieter, "Silk Weaving in Ancient China: From Geometric Figures to Patterns of Pictorial Likeness." *Chinese Science* 12, 1995

Liu, Xinru, *Silk and Religion: An Exploration of Material Life and the Thought of People, AD 600-1200*, Oxford University Press, 1996.

Panday, P.N., S.K. Sharan and P.K. Mishra, *Silk Culture : A Biochemical Approach*, New Delhi, APH, 2005

Sung, Ying-Hsing, *Chinese Technology in the Seventeenth Century - T'ien-kung K'ai-wu*, Translated and annotated by E-tu Zen Sun and Shiou-chuan Sun, Pennsylvania State University Press, 1966.

Index